THE TEACHER'S
3
IDEA BOOK

100
SMALL-GROUP
EXPERIENCES

Michelle Graves

HIGH/SCOPE
PRESS
a division of High/Scope Educational Research Foundation

Published by
High/Scope® Press

A Division of the
High/Scope Educational Research Foundation
600 North River Street
Ypsilanti, Michigan 48198-2898
(734)485-2000, FAX (734)485-0704

Nancy Altman Brickman, High/Scope Press Editor
Margaret FitzGerald, Cover and Text Design
Gregory Fox, Photography

Library of Congress Cataloging-in-Publication Data

Graves, Michelle, 1952
    The teacher's idea book.

    Bibliography: p.
    1. Education, Preschool--Curricula--Handbooks,
manuals, etc. 2. Education, Preschool--Handbooks,
manuals, etc. 3. Day care centers--Handbooks, manuals,
etc. I. Title
LB1140.4G73 1989   373.19'0202   88-35794
ISBN 1-57379-029-X

Printed in the United States of America
10 9 8 7

20.65

Other Titles in the Series

The Teacher's Idea Book 1: Planning Around the Key Experiences

The Teacher's Idea Book 2: Planning Around Children's Interests

Related High/Scope Press Preschool Publications

Educating Young Children: Active Learning Practices for Preschool and Child Care Programs

A Study Guide to Educating Young Children: Exercises for Adult Learners

Supporting Young Learners 1: Ideas for Preschool and Day Care Providers

Supporting Young Learners 2: Ideas for Child Care Providers and Teachers

Getting Started: Materials and Equipment for Active Learning Preschools

High/Scope Extensions: Newsletter of the High/Scope Curriculum

High/Scope Buyer's Guide to Children's Software, 11th Ed.

Adult-Child Interactions: Forming Partnerships With Children (video)

Drawing and Painting: Ways to Support Young Artists (video)

How Adults Support Children at Planning Time (video)

How Adults Support Children at Work Time (video)

How Adults Support Children at Recall Time (video)

Available From

High/Scope Press
600 North River Street, Ypsilanti, Michigan 48198-2898
313/485-2000, fax 313/485-0704

Contents

▌▌ Small-Group-Time Plans Originating From New Materials 71

▌▌▌ Small-Group-Time Plans Originating From High/Scope Preschool Key Experiences 123

IV Small-Group-Time Plans Originating From Community Experiences 175

Preface

The idea for this book originated when I attended a workshop on developing small-group times at the 1996 High/Scope Registry Conference. The room was overflowing with early childhood educators eager for ideas on how to plan and conduct small-group experiences attuned to the needs and interests of a wide range of children. Many thanks go to High/Scope educational consultant Linda Weikel, who led the session and provided participants with hands-on materials, a packet of ideas, and a chance to discuss their successes and concerns. Their reaction to the information made it clear that teachers and others working in child care and preschool programs would greatly appreciate additional small-group-time ideas.

Many others have provided encouragement and support as I worked on this book. Carol Markley and Nancy Vogel willingly tried out ideas in High/Scope's Demonstration Preschool, offered feedback on how children reacted to the materials, and let Gregory Fox's talented photo eyes capture the excitement generated by the children. David Weikart, Mary Hohmann, Carol Markley, Sue Terdan, and Philip Hawkins offered valuable suggestions and probing questions during the book's draft stage. Their input helped greatly to clarify and enrich the content. Finally, Nancy Altman Brickman once more went far beyond the role of editor with her magic for making thoughts come alive in words and pictures that tell the story of active small-group times.

— M. G.

Introduction
Planning Successful Small-Group Experiences

When I first came to work at the High/Scope Demonstration Preschool as a graduate student in 1978, I had been teaching young children in public school and child care settings. Most of my experiences had been with play-based curricula, so the transition to a High/Scope learning environment, which is organized to promote children's plans and initiatives, seemed natural to me. Gradually, however, I began to realize that my behavior and interactions with children during the work time segment of the day differed dramatically from those at small-group time. During work time I was quite comfortable taking a supportive role: I moved freely around the room, playing alongside children, taking my cues from them about how to enter their activities, and occasionally introducing new ideas and materials to extend their play. However, during small-group time I became structured and directive in my interactions with children. It was as if I felt a responsibility to "instruct" children during this part of the day; I felt the need to control children's behavior so they could "learn something."

A memorable example of this thinking occurred on a day that I was being observed by my supervisor in the graduate program. I knew that I needed to demonstrate my understanding of the ingredients of active learning (materials, manipulation, choice, language from the children, and support from the adults), so I planned what I felt was a "sure-fire" experience for the children. I came to small-group time prepared with the following materials: peanuts in the shell, a blender, Saltine crackers, napkins, and small plastic knives.

As the group began, I gave each child a small collection of peanuts. Most of the children had never seen peanuts, in or out of the shell, so they were naturally curious about the shapes, textures, tastes, and smells of the peanuts, and the sounds they made when they were cracked open. However, *my plan* was to give them the experience of making peanut butter, and since we only had 15 minutes, I shifted into my "directive teacher" mode. Though children were still interested in exploring the peanuts, I suddenly swooped the peanuts away and tossed them into the blender (I felt I had to do this because they did not understand that you only use the inside of the peanut, not the whole shell, in making peanut butter). Since we only had one blender and nine pairs of hands waiting to press a single button, there was lots of pushing and yelling, and some children even cried because the loud whir of the blender

At small-group time, children use similar sets of materials for very different purposes. One child makes a birthday cake with Play-Doh balls, straws, and small sticks; another child blows on her straw to see what happens to the Play-Doh.

frightened them. I breathed a momentary sigh of relief when the peanut butter was finally finished, because I knew then that at least the children's waiting was over—it was now time to give each child his or her own crackers and knife. However, I hadn't anticipated the confusion and chaos that would occur as children, inexperienced in using knives, tried to spread peanut butter onto the crackers. In most cases, children's awkward efforts resulted in the immediate crumbling of the crackers. After things calmed a bit, I tried to regain my status as a teacher by making the comment, "So, now you'll know where peanut butter comes from." To add insult to injury for me, one of the children immediately responded with, "Yes, it comes from a jar you buy at the store."

When I discussed this experience with my supervisor, who had quietly taken pages of notes during the small-group time, we concluded that I had tried to provide far too many materials and experiences for one small-group time. In the words of my supervisor, "Just giving each child a handful of peanuts would have been enough." For me, this was the beginning of a process of learning what constitutes a truly active small-group time for children and how adults can help to bring about such experiences. In the years since, as both a teacher and a trainer, I have come to understand that the very issues I had struggled with in the peanut butter activity are common to others who work with small groups of young children. Below is a brief sampling of comments about small-group time that I have collected from early childhood educators attending conferences and inservices:

- "While I do believe in active learning and materials for each child, I also think small-group is a good time for children to learn how to share and listen to their teacher."

- "It's impossible to control the children's behavior unless I tell them exactly where to sit and how much tape they can use. If I don't, I always end up with a big mess that they never want to clean up, and besides, it's very wasteful."

- "The part I don't understand is how children, who are so different in their interests and abilities, can use similar sets of materials and still be productive. I feel like I need to have in mind a specific end-result or outcome for the activity."

This book is an effort to find answers to concerns like these. It describes how to make small-group time a meaningful, active, and engaging learning experience for both children and adults. This introductory chapter outlines guiding principles for successful small groups, discusses the mechanical aspects of small-group times, explains how adults find and develop ideas for small-group experiences, and examines how adults support children's explorations during the small-group activity. The discussion in this chapter is illustrated with actual classroom examples from the small-group activities described in the rest of the book.

Guiding Principles for Planning Successful Small-Group Times

Simply stated, small-group time is a segment of the daily routine in which a group of five to ten children meet with an adult, at a consistent time and place, to work with materials selected by the adult. During small-group time children experiment, explore, create, solve problems, or build with the materials in their own individual ways. As children work, adults support them in pursuing their ideas and observe and listen to the ways children play and converse.

Here are some basic principles for planning successful small-group times:

As you choose learning materials for small-group experiences, consider what you know about the interests and developing abilities of the children in your classroom. For example, around Thanksgiving it might occur to a teacher to plan a small-group time relating to the Pilgrims sharing a meal with the Indians. However, this activity may not make sense to young children since they are not mature enough to understand events in the distant past. This historical event is far removed from their experience. Thus, young children's interest in a holiday like Thanksgiving will most likely focus on the aspects of the holiday they have personally experienced, such as the family gatherings they've attended, the special foods they've eaten, or the hotels or strange beds they've slept in while visiting relatives. To support children's developing understanding of Thanksgiving as a holiday, the teacher might instead plan a small-group time that reflects their life experiences. For example, the small-group activity "Storytelling With Props," p. 198, gives

Children often use lots of consumable materials as they explore them, so plan ways to conserve materials without discouraging exploration. If wasting glue is a concern, try filling the bottles one-third full.

children the opportunity to describe, in their own actions and words, some of the personal meanings the Thanksgiving holiday has for them. It also allows children who are not interested in the holiday to explore the materials in an active way without connecting it to the holiday. Either way, such an activity gives the adult an opportunity to learn something about each child in the group.

To accommodate the needs and interests of children in your small group, choose materials that may be used in a variety of ways. Stay away from items, such as store-bought Lotto games, that involve rules or pieces that must be assembled in the "right way." Instead, select open-ended manipulatives, such as Duplo or Lego blocks, that children can arrange and rearrange, fit together and take apart, or simply dump and fill. Avoid materials for prepackaged art experiences, such as coloring books or cut-along-the-dotted-line paper projects. Instead, offer children art materials, such as markers, pens, pencils, scissors, and paper, that they can use and explore to create their own patterns, designs, and markings.

Have back-up materials ready. A common concern of teachers at small-group time is that younger children may not be able to sustain interest in a particular set of materials for as long a period as the older children in the group—some children may finish after 5 minutes while other children have just begun working on their ideas. Preparing alternate materials that complement the original materials is one way to handle this problem. For example, an adult using the "Noodle Fun" small-group time, p. 30, had markers on hand to offer as a back-up material to children who might lose interest in combining noodles with the glue and cardboard. In many cases, such back-up materials will remain unused throughout the activity, but in this particular case, the adult found that one child tired of arranging and rearranging the noodles on the cardboard shortly after the group started. She also knew that this particular child rarely chose to participate in activities that involved messes. The teacher dealt with this situation by acknowledging that the child had stopped arranging the noodles. The teacher then wondered aloud if the child would like to use the markers with the noodles. The child followed this suggestion, and while he was drawing directly on the noodles, other children in the group were gluing noodles at random or in specific groupings, or using them to make people's faces and bodies. Some adults may worry that such alternate materials will distract children who are working well with the original set of materials, but in our experience, children who are absorbed in using materials usually remain focused.

Use the setting of small-group time as an opportunity to introduce new materials or to revisit old materials the children appear to have lost interest in. During children's explorations observe to see what their actions tell you about the value of the materials and how you might later use them in the classroom. While it is certainly important for teachers to observe children's interests in developing small-group ideas, we don't mean to suggest that every aspect of small-group time should relate to interests that children have explicitly expressed in their play. For example, the teacher who planned "Flashlight Tag," p. 140, had not observed children talking about or using flashlights. It simply made sense to this teacher to add a new material, flashlights, to an activity that children participated in on an everyday basis. Having observed their reactions to the flashlight at small-group time, it then made sense to her to add flashlights to classroom areas as a work time choice and to use them also in planning or recall strategies (asking children to shine a flashlight on something they chose to work with). When you notice children have stopped choosing certain materials at work time, you may want to use small-group time as a way of rekindling their interest. For

Small-group time can be an opportunity to revisit familiar materials (such as toothbrushes) and use them in new ways (to paint a mural on the floor).

example, offering toothpicks and sticks with Play-Doh as described in "Working With Sticks, Straws, and Play-Doh Balls," p. 124, gave the children in one group possibilities they had not considered, and rekindled their interest in Play-Doh as a work time materials option. Having the opportunity to use similar materials on succeeding days and in other parts of the daily routine will remind children of the many possibilities offered by the materials.

As you plan particular small-group activities, consider the time of year, how long the children have been together as a group, and the experiences they've already had with the materials you would be providing. Some of the small-group experiences suggested in this book are designed for early in the school year, while others are more likely to be successful as children mature and get to know the program, their classmates, and the teaching adults. For example, the activity "Making Wood-Scrap Printing Tools," p. 142, involves complicated move-ments and is thus best suited for older preschoolers who have been in the program for some time. However, the "Stone Drawings" activity, p. 164, could be used early in the school year, with younger children, because of the simplicity of the materials used.

Sources of Ideas for Small-Group Times

At a recent High/Scope workshop one participant attending a session on small-group time raised an issue that had a familiar ring to others participating in the group. "Our school plans around themes," she stated, "so if I'm not going to be studying about bugs and insects in June, where will my ideas for small-group time come from?" In the High/Scope Curriculum, we have identified four key sources of ideas for small-group times: **children's interests; new materials; the High/Scope key experiences;** and **children's experiences with the events, places, and traditions of the community.** This book presents 25 sample small-group plans for each idea source. In the following sections, we provide some introductory guidelines for planning each of these four kinds of small-group times.

Planning around children's interests

When adults are in the habit of observing children closely, children's actions often suggest ideas for small-group experiences. For example, adults may observe children using building materials or art equipment in specific ways, developing relationships with adults and peers, imitating the actions of workers they see in the neighborhood of the school, or working out their feelings about exciting or scary events. Any of these observations could be the starting point for a small-group time plan. The small-group time "'Painting' Outdoor Structures," p. 22, illustrates this point. This activity originated when teachers observed children watching a maintenance worker paint the building across from the school. Similarly, "Animal Care," p. 36, was planned when teachers noticed children pretending to be dogs and cats during the work time segment of the daily routine. In this activity, teachers planned for Alex, a staff member's dog, to visit the classroom.

Planning around new materials

It is natural for teachers to want to introduce children to new materials and experiences. A major part of the teacher's role is to be a resource person who presents children with a stimulating and challenging environment within which children can explore and experiment. Exposure to a wide variety of indoor and outdoor materials; physical activities; new tastes, textures, and smells; and opportunities to observe many natural phenomena—these are experiences that help satisfy children's natural curiosity and develop their awareness of the value and importance of differences and change. Small-group time offers an excellent setting for such adult-initiated experiences with new materials. For example, in "Exploring Packing Peanuts," p. 78, children have an opportunity to make discoveries about sinking and floating and how the texture of this unusual material changes with the addition of water. In a similar way, "Pomegranate-Apple Explorations," p. 76, gives children a firsthand experience in comparing two fruits whose outside surfaces are similar, but whose insides are remarkably different. Keep in mind that while most of the small-group times in this category involve tangible materials, teachers may also use small-group time to introduce new experiences in movement and music that do not involve materials in the usual sense. In these cases the "materials" children manipulate at small-group time may be their own bodies or voices, or the taped musical selections they are listening to or moving to.

Planning around the High/Scope key experiences

As team members reflect on their knowledge of individual children, they may notice that there are some areas in which they have little information about a child's abilities. The High/Scope key experiences provide a framework for identifying such areas as well as for planning experiences that will yield such specific information about children. In planning small-group times around specific groups of key experiences, adults will also want to consider observations they've made of children's interests and actions. For example, "Angry Stories," p. 152, is a small-group plan that adults designed with the hope of learning something about individual children's abilities to express their feelings in words (instead of with the hitting and pushing they were observing in the classroom). Similarly, "Super Bubbles," p. 156, was planned after adults observed children's interest in comparing the attributes of the bubbles they were making while splashing around in soapy water during work time. Thus the key experiences provide a backdrop and framework for understanding children's actions and ideas and for providing further experiences that build on these actions.

Planning around community experiences

Within the community where you live and teach, there is a richness of tradition that can be used as a source of small-group time ideas. Some of these ideas may come from the interests and experiences children and their family members bring from outside the classroom, and other ideas may come from the experiences you, as the adult, bring to them from your own home and community life. In either case, capitalizing on the excitement of a local event or a family's participation in a work or community activity will enrich the children's awareness of the world around them. As you read through the section on community experiences in this book, you'll notice that many small-group time ideas relate to seasonal changes, holiday celebrations, or community events such as bike racing, bowling tournaments, and high school fundraisers.

Following this visit to a local 4-H livestock show, teachers planned a small-group time around imitating the sounds of animals.

When reading through these activities we suggest you choose those that seem to fit with the community of children you work with and eliminate those that won't make sense to them.

For example, the activity "Ice Sculptures," p. 204, may not be meaningful to 3-year-old children who live in warm climates. By contrast, the "Neighborhood Treasure Walk" activity, p. 176, can be adapted by the teaching team for most groups of children, with the understanding that the group will see and collect very different things, depending on the locations they are exploring.

The Mechanics of Small-Group Time

This section presents general guidelines for structuring the mechanics of small-group experiences. These techniques are based on our knowledge of children's need for stability, consistency, and active participation.

Choose a consistent time and place for your small group to gather. Spend between 15 and 20 minutes daily with your group. Divide up the children in your class so that each adult works with about the same number of children. For instance, if the class has 18 children and two adults, each adult would have 9 children in their small group. We recommend that you meet daily in the same location as a way to initially gather the children. However, there will be days when the space you typically meet in does not accommodate the plan for the day. For example, though you may normally meet at the art area table, your activity for the day may be planned for the computer area. In such cases, meet first in your regular spot and then move with the group to your planned location. This is recommended because, over time, children will internalize the routine and independently move to the location you usually meet in.

Keep the membership of your group constant for a period of several months at a time. This allows children and adults the opportunity to establish relationships and interaction patterns. Keeping the group constant gives you the chance to closely observe and study individual children in a setting that is smaller and more intimate than the overall classroom setting. You will observe individual children's developmental growth and the growth of the group as a social unit.

Keep the developmental levels of children in mind as you choose materials for small-group time. This teacher had observed that her group of older preschoolers were adept with their hands and enjoyed stringing beads. So she provided materials for another kind of string experience.

Eventually, you will be able to predict the ways children may respond to the materials and experiences you bring to them. When you do make changes in the group's make-up, you may want to consider new groupings that will help you learn about children's relationships outside the group. For example, consider grouping together two children who often make work time plans together or separating twins who always play together.

Have materials ready before the group begins, and keep your introduction to these materials brief. Children who are expected to wait while you gather materials for the group may find a way to fill the time with behaviors that may make it difficult for you to gain their attention when the materials are ready. We recommend that you prepare materials in individual containers (such as paper bags) for each child before the start of the school day, storing them near your small-group meeting spot, but beyond children's reach.

Think beforehand about the individual children in your group and how they may react to the materials. For instance, if you want to introduce Legos, but you know that two of the children in your group may have difficulty manipulating the small pieces, also include a similar but more easily manipulated material, such as the larger Duplo blocks. This kind of planning helps to prevent frustration, both for children and adults.

Have available a safe and consistent alternative activity for children who do not want to participate in small-group time. For many different reasons, on some days you may find that certain children do not want to join in the activity. Offering these children a quiet place in which they can look at books or color, for example, is an alternative that keeps them occupied and prevents them from distracting others in the group who are interested in the planned activity. Make sure the location you have chosen for this activity is within your line of vision.

If you have followed the recommended strategies to prepare for small-group time, you will be ready to take the next step—**focusing your energies on maintaining an active learning experience for all children.** This is a challenge that requires a supportive interaction style that will be described more fully in an upcoming section. Later, when small-group time has ended and the children have left for the day, spend a few minutes examining what happened. Your reflections on the children's actions and language will help you determine whether you maintained a good balance between providing a structure yet allowing children the freedom to explore in ways that were meaningful to them.

The Small-Group-Time Planning Form

Once you have decided on an idea for a small-group activity, the next step is thinking about and recording your ideas. As you plan small-group experiences, it is helpful to have a system in place for considering in advance what children might do or how they might use the materials you provide. This kind of pro-active thinking and planning will help you resist any impulses you may have to control the direction the group takes.

A helpful tool in this planning process is the form we use for the small-group activities in this book. A blank copy of this form is presented on the next page. Since this form is lengthy, some teachers may not find it practical to fill out completely on a daily basis. However, the form is presented here because it reflects a thinking process that we feel is necessary for planning effective activities, even if teachers choose not to record their plans in such detail. Teaching teams are encouraged to adapt the form to their own program needs. Each section of the form is discussed next.

Small-Group-Time Planning Form

Originating ideas

Possible key experiences

Materials

Beginning

Middle

End

Follow-up

Sections of the Form

Originating ideas

As you observe children's actions and interactions in the classroom, something about these behaviors and events may spark an idea for a small-group experience. Whether the idea grows from a child's interest, a new material you want to introduce, a local event, a key experience area you would like to highlight, or some combination of these, you can record your ideas in this section of the planning form. Doing so will help you assess whether you are providing a balance of experiences at small-group time.

Possible key experiences

When children are actively involved in small-group-time experiences, they will manipulate the materials provided in a variety of ways, talk about their discoveries rather than listen to adult explanations, encounter and solve problems, and spontaneously discuss additional materials they need to make their ideas work. As they do so, they will naturally encounter many of the High/Scope key experiences. The ways adults observe, interact with, and enter children's conversations will support these ongoing opportunities for learning. As adults plan, the possibilities inherent in the materials and their knowledge of the children in their group should help them make predictions about what key experiences may occur during the small-group time. Stating these on the form will help teachers develop skill in anticipating and planning ways to support the development of important skills and capacities.

Materials

Use this section to make a list of the materials you will need to prepare for the small-group time. Materials listed should usually include a set of materials for each child to work with; additional materials needed by the teacher to introduce or carry out an idea, including books or recorded music; containers for storing materials or for use in cleanup; recipes needed to make materials; and back-up materials to have available to use as needed.

Beginning

This section records ideas for opening the group activity and predictions of how children may respond. Children usually arrive at the small-group space eager to begin work. Since we know that what they do with the materials provided will vary according to their interests and developmental levels, we suggest that introductions be kept brief and open-ended, so children can get started quickly and will not feel obligated to create something suggested by the adult. The ideas listed in this section in the small-group plans given in this book will give you some ideas for simple statements or game-like ideas that will call attention to the materials and help children put their own ideas in motion.

Middle

In this middle section you can record ideas for how you might interact with children as they work. You may also use this space to describe some of the wide range of behaviors you might

The notes you write on the middle of the small-group-time planning form will help you prepare for the many different ways children may use materials. Here Jake devises a new painting technique, first carefully painting his hand with a toothbrush, then making a hand print.

expect from children as they approach materials. Thinking through your expectations about children's behavior is especially important because it will help you develop strategies in advance for responding to both developmental and interest differences among children. If you have prepared in this way, it will not seem surprising, for example, if some of the children simply pound and squeeze the Play-Doh pieces you have provided for an activity, while others roll them into the shapes of letters and numbers. For a child who is pounding and squeezing Play-Doh, the adult may plan to provide support by imitating and labeling the action and repeating and responding to any language the child produces. If a child is making numbers out of Play-Doh, on the other hand, the adult might want to work alongside the child, making small groupings of Play-Doh balls that match the numbers produced by the child. This process of thinking through possible support strategies in advance will help you see the learning value of both sorts of activities and make it less likely that you will try to place more importance on one action than another.

End

Not all children can be expected to finish working at exactly the same time. This section of the form will help you think through ways to bring closure to the small-group activity so some children can finish while others are still working. You can also use this section to list strategies for cleanup and materials storage, as well as tips for including children in the process of deciding where materials introduced at small-group time may be stored in the classroom for use at work time.

Follow-up

In the High/Scope Curriculum learning is not viewed as happening on a "one-shot" basis; instead it is seen as occurring over long periods of time, as similar experiences are repeated.

Thus, in the follow-up section of the planning form, you should list options for expanding on the ideas children have encountered in the small-group activity. Some of these ideas may involve adding materials to the existing room arrangement, while others involve developing future strategies for other parts of your daily routine, such as planning, recall, large-group, or outside times. The planning section may also contain additional small-group time ideas that are related to the original activity. For example, a teacher may plan a series of small-group times in which children can explore painting on various surfaces.

In summary, the small-group-time planning form is designed to help adults think through the planning process: the sources of ideas, what materials can best foster growth and challenges for children, and how these materials might be used in different ways by different children, without compromising the quality of the learning opportunity. The next section discusses adult support strategies for conducting small-group time.

Supporting Children During Their Small-Group Explorations

Building on children's individual discoveries during small-group time is a challenging task that will keep adults busy and active throughout small-group time. The following are strategies adults use to support children's learning opportunities as the small-group time unfolds.

Stay on the children's physical level. Some small groups are held at a table, others may be held around a large piece of paper on the floor, and others, outdoors on the playground. In any case, adults should position themselves as children do. If children are working on the floor, getting down at their level helps you understand their unique perspectives and sends them the message that you are ready and available to join them in their excitement. For example, when conducting the "Neighborhood Treasure Walk" activity, p. 176, a teacher joined the children in crouching down to watch a colony of ants moving bread crumbs into a small hole in the ground. As she knelt on the sidewalk with children, she overheard a passerby say to her friend, "You know, it takes a child to help you remember how fascinating ants can be."

Observe the things that children do with the materials. Small-group time will always be full of surprises because children will have a wide array of responses to the materials presented. For adults, this opportunity to closely observe the very different ways children work with similar sets of materials provides a mini-laboratory for learning about their interests and skill levels. For example, as one class participated in the small-group activity "Shape Template Drawings," p. 126, children used the materials and described their actions in their own ways. Tamara used several different templates and told her teacher she had made "Rudolph the Red-Nosed Reindeer." Christopher, who had connected his shapes in what appeared to be a random pattern, stepped back from his work and said, "This looks like a bicycle." Jake spent the entire small-group time experimenting with tracing around the different shapes available, being careful to position each new one so it did not touch the edges of any that he had already traced. An adult who observes such activities with a trained eye can learn a great deal about children's fine-motor skills, representational abilities, and spatial awareness. The adult can use these observations in planning ways to support individual children, for example, adding a book about reindeer to the book area for Tamara to use, examining a real bicycle with Christopher, or providing opportunities for Jake to arrange and rearrange shapes in enclosed spaces.

Carefully listen to what children say. As they work with small-group materials, children may make comments to the teacher, to other children, or to themselves that yield many insights about their understanding of the world around them. As children participated in the "Toothbrush Mural" activity, p. 32, for example, the adult overheard a revealing conversation between Jake and Reid. Reid approached Jake first, saying "Can you give me that yellow so I can make green?" (He already had a cup of blue paint.) After Jake passed him the yellow paint, he watched as Reid made green paint by mixing the yellow and blue paints. Then Jake said, "Now pass it back to me so I can do it, too." Reid reacted to this with, "I'm not done yet, and I probably won't be done for a long time." Jake then shouted back, "Now that doesn't seem fair!" after which Reid smiled and handed him the paint. To a trained adult this exchange reveals a great deal about Reid's and Jake's ability to interact socially and to "get through the rough spots" by using negotiation instead of force without involving an adult.

Give attention to each child. As teachers, you know that there are days when children call for our attention 10 times over a 5-minute period, and other days when the same children are so absorbed in what they're doing that they don't notice we are right next to them. By moving around from child to child at some point during the small-group time, you will gain an understanding of what each child is doing and saying, and you will avoid missing the quiet child who doesn't ask for your attention. Children will soon begin to understand that you can be relied upon to set aside time for them during the small group, freeing them to focus their energies on their work.

Do the same things children do. Imitating a child's actions is one way to show you recognize and strongly support his or her actions. Children who are involved in activities are delighted when adults join in to share the excitement. One day at small-group time, children were doing the "Paper-Plate Streamer Dancing" activity, p. 172. As they moved to the music, waving their paper plates festooned with streamers, Jota's father, who had come early to pick him up, stepped quietly into the large-group space. He picked up an extra plate with streamers, and he began moving his arms and feet in time to the music. Jota stopped as though frozen. Looking first at his father and then at his classmates, Jota said, "Look, my daddy is doing it too." Thus, when you imitate children's actions you become a partner in their explorations without changing the direction of their chosen activity.

Have conversations with children, being sure to let them lead the way. Often children who are focusing on their work enjoy a quiet atmosphere, and a comment or question from you may interrupt their thoughts. Just as often, however, children will want you to notice, acknowledge, or talk to them about what they are doing. In these cases, take your cues from the child so the conversation stays focused on what is of interest to the child. For example, during the "Sun Reflections," small-group time, p. 34, Audie was jumping up and down with his mirror in his hand, saying "Look, look, look, my sun is dancing." Trying to match the enthusiasm in his voice, the teacher said, "When you jump up and down, you can make your sun dance." After spending a few minutes with Audie, the teacher sat down next to Emma, who had spent most of the group time by herself, sitting in one spot on the floor and moving her sun reflection up and down the wall from the floor to the ceiling. Her teacher acknowledged this by watching silently for a few minutes, and then saying "Down on the floor, up to the ceiling." When Emma did not respond, the teacher accepted her silence and moved on to another child. The teacher knew that there would be other opportunities to interact with Emma during future activities.

Resist the urge to do things for children. Instead, encourage children to do things on their own or refer them to other children who may be of assistance. For example, while participating in the "Nuts and Bolts" activity, p. 48, Mark tried several times to put a small nut on a too-large bolt. Observing this, the teacher was tempted to give him the correct-sized bolt because she knew that when materials frustrated Mark, he often reacted by throwing things, crying, or hitting others who were nearby. However, she resisted this temptation, instead commenting to Mark that it was hard when the two things you used weren't fitting together. She then referred him to another child who was close by: "Mark, I noticed Katie using a smaller nut; maybe you could ask her to pass you one like she has to try."

Ask questions sparingly. Too often, well-meaning adults disrupt children's activities by asking too many questions. While the adult's purpose may be to gain information about a child's understanding of something, the child may experience such questions as ill-timed and unwelcome intrusions. For example, during a color-mixing activity an adult might interrupt a child's explorations with the question, "What colors made that purple on your paper?" (even though the child is now working with red and white paint). By contrast, in the High/Scope approach, we suggest that teachers usually refrain from questioning children, because we believe that, to children, the most relevant questions are the ones they ask themselves. Instead of posing questions, High/Scope teachers often make acknowledging comments related to children's ongoing work. Occasionally, situations arise in which children stop to consider a problem. At these times it may be appropriate for the teacher to ask a question that presents a challenge to the child's thinking. "I need pink, I need pink, I need pink," chanted Jacob during the "Toothbrush Mural" activity, p. 32. "Pink is a favorite color of yours,"

Silent watching may be the best kind of support for a child who is concentrating on his work. On the other hand, a child who is getting ready to put his work away may enjoy conversing about it.

acknowledged his teacher. "Yes, and I need pink right now," he said. This was an opening for the teacher to ask a question that might open his eyes to a possible solution: "I wonder what would happen if you mixed two of the colors together?" Jacob did take up the teacher's implied suggestion, although he didn't immediately succeed in getting pink because he first tried mixing green with yellow. Eventually, with the help of a classmate who suggested he try mixing red and white, he made the pink color he hadn't been able to find in the paint cups.

Stay flexible as you conduct small-group activities, allowing children to explore materials in their own ways and following their cues about the direction of the activity. At the same time, be sure to establish safe and reasonable limits for their behaviors and actions. For example, the activity "Finger Printing With Sponges," p. 72, allows children to choose how, and with what, they will make prints on paper. The teacher provides an assortment of materials for making prints on paper. The directions for this activity suggest that the teacher initially encourage children to paint with their fingers, but teachers must also be aware that children are likely to try painting with their knees, elbows, and other body parts. While a teacher would probably encourage this kind of experimentation, what if a teacher observed a child printing on another child's clothing, running through the room with paint dripping from her fingers, or attempting to paint-print with the pegs from the toy area or the mouse from the computer area? In such cases the teacher would select words to help children understand that there are limits to the kinds of explorations possible: "Asia, I'm going to stop your hands from painting on the computer mouse, and I'm wondering if you can guess why." Including the child in the discussion about why those limits are needed will help the child accept the limit and will lead to a deeper understanding of the situation than would result from simply stopping the behavior without explanation. On the other hand, it's important to anticipate that children will often use materials in an exploratory way before they begin to use them in conventional ways. Since waste is a concern in all early childhood programs, you'll need to anticipate and plan for children's explorations and, at the same time, plan ways to conserve materials. If you are planning an activity with glue, for example, you will expect that children will at first explore the glue bottles, take off the lids, and possibly turn the bottles upside down before they begin to glue things on paper. Therefore, provide a glue bottle for each child in the small group, but fill each bottle only one-fourth full.

Help children make the transition from the end of small-group time to the beginning of the next part of your daily routine. Providing children with support for their ideas means realizing that everyone will not finish at exactly the same time. Help them make a smooth transition by offering a safe, yet stimulating option for those children who finish sooner and by supporting those children who are not finished as they think through ways to complete their ideas. For example, a teacher who tried the "Fabric Banner Painting" activity, p. 62, found that washing out paintbrushes and hanging the banners with clothespins kept the early finishers busy. With these children occupied, she was free to talk to the children who were still painting about finding a place to hang their banners so they could complete them at the next day's work time.

No matter how creative your plans for small-group time, the success of your group experiences depends on your ability to support children's ideas. This means remaining flexible and taking your cues from children as they manipulate the materials. It means understanding that young children *will* learn, even when you don't "instruct" them in prescribed ways, with your own agenda in hand. The next four sections of this book present many exciting ideas for small-group times that originate from the **children's interests; new materials; the High/Scope key experiences;** and **community experiences.** These small-group plans are not intended to be used as blueprints for activities in your program. Instead, we encourage you to use the processes described in this chapter to adapt the plans for your unique group of children. Then, as each small-group time unfolds, remember to be receptive to children's initiatives, interests, and plans.

I

Small-Group-Time Plans Originating From

Children's Interests

Stacking, Balancing, and Nesting Boxes

Originating ideas

Since the addition of the smaller cardboard bricks to the block area, children have been making stacks of big and small bricks, often talking beforehand about putting the big ones on the bottom so the pile is "sturdy."

Possible key experiences

Language and literacy: *Describing objects, events, and relations*

Initiative and social relations: *Solving problems encountered in play*

Seriation: *Arranging several things one after another in a series or pattern and describing the relationships (big/bigger/biggest, red/blue/red/blue)*

Space: *Fitting things together and taking them apart*

Materials

✔ A collection of **different-sized boxes,** such as those from food, jewelry, or clothing (Be sure there are enough so each child can have at least five or six.)

✔ For back-up: **masking tape**

Beginning

Bring to the table a big box that contains several smaller boxes, nested from smallest to largest. Tell children that the box has something inside. Pass it to one child and see if that child can guess the contents by shaking it. When the child is finished guessing, have him or her open the box to discover the next box. Repeat this procedure, with other children taking turns at guessing and shaking, until all the boxes have been revealed.

Middle

Take the remainder of the boxes to an open floor space, and work alongside children—stacking, building, and balancing the boxes. Move around the group, and comment on the ways children are using the boxes. For example, if they are fitting them together, as in the beginning activity, comment on their actions ("You put a tiny box inside a huge one"). Make suggestions that relate to their explorations ("I notice you have a really big box—I wonder if you can find two boxes on the floor that will fit inside this one"). Offer suggestions for solving problems ("I wonder what might happen if you turn the box sideways—it might fit then"). When you see children fitting boxes together using trial and error, acknowledge their efforts ("You're working hard to squeeze that big box into the smaller one"). Observe to see whether any children demonstrate more mature thinking, for example, accurately predicting—without first experimenting—which boxes will fit inside other boxes. Listen for the language children use in describing their actions, noting whether they use spatial language ("under," "behind," "in front of") to describe the positions of their boxes. Challenge their thinking about their descriptions with statements like this: "From where you're sitting, the big box is behind the little one. Look what happens when you come over to where I'm sitting."

End

Give the children a warning when there are a few minutes left of box play. As a cleanup activity, make a game of trying to fit as many of the smaller boxes as possible into the larger boxes.

Follow-up

1. Add the boxes to the block area to see if children incorporate them in their building plans.

2. Bring the boxes to planning or recall time. Ask children, in turn, to choose a box and to find an object that fits inside that they plan to use at work time (or did use). Then have them talk about their plans for, or experiences with, that item.

"Painting" Outdoor Structures

Originating ideas

For several days children have been watching a house painter working on a building across from the school.

Possible key experiences

Creative representation: *Pretending and role playing*

Movement: *Moving in nonlocomotor ways (anchored movement—bending, twisting, rocking, swinging one's arms)* and *moving with objects*

Materials

✔ An assortment of **different-sized brushes** (watercolor brushes, brushes for easel painting, housepaint brushes)

✔ **Coffee cans, pails, or other containers,** one for each child

✔ **Water**

✔ **Paper or cloth towels**

✔ For back-up: **paint rollers** of various widths and **paint trays**

Beginning

Give each child a coffee can or pail to fill with water and carry outside. Gather near a place where children can "paint" the outside of the school building and the pavement nearby. Show children the paintbrushes and rollers and say "Let's see what happens if you use these materials to paint the school, the way the house painter has been doing."

When appropriate, involve children in the preparation of materials for the activity. Here, Adam gets a bucket of water to carry outdoors for a water painting project.

Cirra and Stone pretend to be house painters after watching a worker paint a building across from the school.

Middle

Note the ways children fill and carry their water containers outside, observing which children can carry independently, which seek out other children or adults to help, and which refill containers on their own if they spill as they are being carried. As they begin "painting" on outdoor surfaces, see which children imitate the painting actions of the house painter and which children make other kinds of motions with their brushes. Talk to them about these differences ("When you move your arm up and down like that, you look like a house painter" or "You're making dabs of water on the sidewalk—the house painter we've been watching has been painting in a different way"). Imitate children's actions and pause to listen to their comments about and descriptions of what they are doing.

End

Provide paper or cloth towels and ask children to squeeze their paintbrush bristles into the towels. Find a sunny spot on the playground where children can empty the water from the cans. Return brushes and cans to the classroom.

Follow-up

1. Later in the day, go outside with children to see what happened to the wet areas of the building and playground.
2. Bring brushes and water back outside and make them available as a choice for outside time.
3. Repeat this small-group time again in different weather (during a light drizzle, on an overcast day, on a bright, sunny day). Encourage children to compare the two experiences.

Be available to support children's spontaneous discoveries during small-group time, as this adult was when children painting the basketball hoop with water noticed a spider climbing up the pole.

Autumn Leaf and Flower Prints

Originating ideas

This activity was planned in early fall. Teachers had observed that children were becoming aware of various seasonal changes: some children had been playing with fallen leaves on the playground; others had noticed that the flowers growing in the window boxes outside the classroom were starting to wilt; and others had watched and imitated the school maintenance person as she prepared the flower beds for winter and dug up some of the plants for composting.

Possible key experiences

Creative representation: *Relating models, pictures, and photographs to real places and things*

Language and literacy: *Describing objects, events, and relations*

Movement: *Moving in nonlocomotor ways (anchored movement—bending, twisting, rocking, swinging one's arms)*

Classification: *Distinguishing and describing shapes*

Materials

✔ Several varieties of **flowers** gathered from school beds or window boxes, enough so that each child can have several blossoms or sprigs of each variety (Marigolds, impatiens, and yellow daisies are common flowers that will work for this activity—blue lobelia works especially well if you can locate some.)

✔ **A large piece of unbleached muslin** or individual pieces of **plain white paper**

✔ **Hammers** or other pounding tools, one per child

✔ **Paper bags** to cover the table or floor

✔ For back-up: **markers or crayons** (with wrappers removed)

Beginning

Cover a smooth, hard surface with paper bags, then lay the sheets of paper or the muslin piece on top. Give each child a collection of flowers. Tell children that they can use their flowers to make a design on their piece of paper or a part of the cloth. When they have finished arranging their flowers, encourage them to help you as you place another large cloth or sheets of paper on top of their designs.

Middle

Suggest that children run their hands over the paper or cloth to see if they can feel the leaves and flowers underneath. Then pass out the hammers or other pounding tools, and ask them what they think will happen if they hammer in the places where the leaves and flowers are. Acknowledge their responses by repeating their words ("You think the table will crack" or "You think the flowers will break"). Watch as they discover that the pounding motions bring out a print of the flowers or leaves underneath the paper or cloth. Listen for comments about the shapes or colors children recognize as they bleed through the fabric or paper, and encourage children to search for more flowers as they pound ("You *did* find a yellow daisy—I wonder if you can find a yellow marigold"). Watch to see if any children rearrange their designs to include different colors or shapes they have seen other children using or creating.

End

As small-group time comes to an end, enlist children's ideas on where they can display their flower-decorated cloth or papers. Scrape excess flowers and leaves off the paper or cloth and dispose of them in your compost pile, if you have one.

Follow-up

1. Add the materials to the art area as an option for future work time plans.
2. If your group made a cooperative mural, the day after the small-group time, hang it up on a low window or wall. At greeting time, point out the picture to children, telling them that you plan to give it to the school maintenance person as a thank-you for her hard work in the gardens. Explain that they can make a work time plan to "write" her thank-you messages to go with the picture or to have you write their messages for them.

Scary Eyeglasses

Originating ideas

For the last two days Bryan has been coming to greeting circle with a pair of cardboard eyeglasses (a toy that came with a child's meal at a fast food restaurant) on his face. He makes scary noises, the other children laugh, then he takes off his glasses and says, "Oh, don't worry, it's just me—Bryan—under the glasses." When other children ask Bryan if they can wear his glasses he says no, telling the children to go to the same restaurant to get their own glasses.

Possible key experiences

Creative representation: *Pretending and role playing*

Initiative and social relations: *Solving problems encountered in play*

Movement: *Moving in nonlocomotor ways (anchored movement—bending, twisting, rocking, swinging one's arms)*

Space: *Changing the shape and arrangement of objects (wrapping, twisting, stretching, stacking, enclosing)*

Materials

✔ A pair of **eyeglass frames** for each child (Local optometrists' offices will often donate frames whose styles have become outdated.)

✔ **Decorating materials** for the frames: for example, short bits of **curly ribbons, glue, sequins, glitter,** and **clip-on barrettes**

✔ For back-up: three or four **hand mirrors**

Beginning

Bring materials to the table and say something like, "Today for small-group time I brought eyeglasses for each of you, along with materials for making them colorful and fancy like Bryan's. See what you can do with them."

Middle

Since this is a brand-new material for children, begin by decorating your own pair of eyeglasses. Use the ribbons for tying and twisting, and watch as children decide which materials to use and solve the problem of attaching things to the eyeglasses. Talk to children about their choices ("I notice you only used blue barrettes") or solutions to problems ("When you asked Tracy to hold the glasses for you, it seemed easier to tie on the ribbons"). Notice which children stay absorbed in the decorating, and which use the eyeglasses without decorating them, putting them on and taking them off. Play along with children as they create games and pretend with the glasses, accepting the roles they assign you. For instance, if someone puts on glasses and says "I'm scary" to you, act scared and back away from the person.

End

After the decorating materials are all cleaned up, ask children if they want to keep wearing their glasses or if they would prefer to put them away in their cubbies. Store additional eyeglasses near the large-group-time space, so children can use them to have an eyeglass dance or a parade to music.

Follow-up

1. Use an additional pair of eyeglasses at planning or recall time. The child wearing the glasses looks toward the interest area where he or she plans to play (or did play), then tells the others his or her plans or experiences.
2. Add eyeglasses to the dress-up materials in the house area.

Slime

Originating ideas

Recently during work time, children have shown a strong interest in squeezing and pouring glue, while showing little interest in using the glue as a fastener. Three children in particular have been fascinated by the movement of the glue, using phrases like "It's running" and "Look how high it falls" to describe what the glue is doing.

Possible key experiences

Language and literacy: *Talking with others about personally meaningful experiences*

Movement: *Describing movement*

Classification: *Exploring and describing similarities, differences, and the attributes of things*

Space: *Filling and emptying*

Materials

✔ A **water table** or individual, large **plastic food storage containers or bowls**

✔ **Slime-making ingredients—water, cornstarch, and food coloring** (Slime is a pourable, semisolid material made by combining one part water, tinted with a touch of food coloring, with four parts cornstarch.)

✔ **Smocks**

✔ **Plastic sheeting** to cover the carpet below the table (You don't need this if you have a tile floor.)

✔ For back-up: **plastic tubes, spoons,** and **spatulas**

Noticing children's fascination with the texture of glue, this teacher prepared a less costly alternative—slime—for children to explore.

Beginning

Have children put on their smocks. Gather around the containers or the water table, and give each child a container of premeasured water or cornstarch to pour into the table or bowl. Listen to the ways the children describe the changes as the colored water and cornstarch are mixed together. When the ingredients are completely mixed, the slime will be milky in color. It will look like a liquid but will have a thick, smooth consistency. The slime will thicken further when you touch it with your hands, although it will still be very soft and slippery and will run out of your hands if you try to pick it up.

Middle

Begin working the ingredients with your hands, inviting children to join in. As you and the children continue to manipulate the material, listen for children's comments. Note whether they use spatial terms, such as "up," "down," "into," or "out of," and repeat and acknowledge what they say ("It does slip down through your fingers when you hold your hands up"). Ask children to make predictions ("Do you think the same thing will happen if you hold your hands sideways?").

Children discover some of the unique properties of slime as they use it with the tools the teacher has prepared beforehand as backup materials for the activity.

End

Allow ample time for cleanup. Soap and water will remove slime from children's hands. Don't wipe off the tile floor or plastic sheeting; instead encourage children to look at it carefully. With the children, check it again in about 30 minutes. By then, the slime will have changed into a hardened, powdery substance that can be easily vacuumed.

Follow-up

1. Leave the slime in the sand and water table or in large containers to be used at future work times. It will harden but can be restored to its original form by adding a little bit of water.

2. Use slime in a new way outdoors in a small-group time or outside time a few days later (see "Outdoor Slime Racing," p. 146).

Noodle Fun

Originating ideas

At a recent lunch time, there was a lot of conversation about the different noodle dishes two of the children brought in their lunches. One child brought chow mein noodles and the other brought sesame noodles. Conversation focused mostly on the texture of the food (crunchy versus slippery).

Possible key experiences

Classification: *Exploring and describing similarities, differences, and the attributes of things* and *sorting and matching*

Seriation: *Comparing attributes (longer/shorter, bigger/smaller)*

Materials

✔ **Cooked and uncooked noodles** in a variety of shapes, such as wagon wheels, spirals, and bow ties (Substitute another material if you are not comfortable using food as a play material.)

✔ **Cardboard pieces,** one per child

✔ Bottles of **glue,** one per child

✔ For back-up: **markers**

Beginning

Give each child a container holding a variety of noodles. Tell children, "Today I brought each of you some noodles. See what you can notice about them."

Middle

Observe children's explorations and be available to support and acknowledge their discoveries. Some children will focus on the different shapes of the noodles, making piles of matching shapes. Others will focus on the texture differences. Use language that acknowledges, responds to, or extends upon what children say. For example, if a child says, "This one is squishy," you can respond by simply repeating what the child says ("It *is* squishy"). Another possible response, after you repeat the child's observation, is to make a comment that raises a question and encourages further thinking ("It *is* squishy. I wonder how it got so soft!"). Accept children's responses ("The oven did it"; "The store made it that way"). If children seem to be losing interest in exploring the noodles, bring the cardboard pieces and glue to the table. Notice the ways children attach the noodles to the cardboard. Do they organize the noodles into groups, glue them randomly, or use them to make something to which they attach a specific label ("It's a person with a bow tie")?

These pasta people—made with Play-Doh, straws, and dried noodles—were created during a follow-up small-group time in which teachers offered some additional materials with noodles.

End

Work with children to collect any leftover noodles, sorting the cooked and uncooked noodles into separate containers. Ask children to tighten the tops of the glue bottles and to wipe off the table. Ask children to store their artwork where it can dry.

Follow-up

1. Fill the sand and water table with a mixture of cooked and uncooked spaghetti.

2. Bring children's artwork back the day after the activity, and provide children with paint and brushes for painting their creations.

3. Add the book *Cloudy With a Chance of Meatballs* by Judi Barrett to the book area, and after reading it to children, note whether they incorporate any elements of the story in their pretend play at the noodle-filled table.

Toothbrush Mural

Originating ideas

After new toothbrushes were introduced in the beginning of the school year, Cory asked what would happen to the old, used brushes.

Possible key experiences

Creative representation: *Drawing and painting*

Classification: *Using and describing something in several ways*

Initiative and social relations: *Solving problems encountered in play*

Materials

✔ Used **toothbrushes**

✔ A long sheet of **paper** (Make the paper wide enough so children seated on both sides will have room to paint, but narrow enough so that children can easily reach the middle of the paper.)

✔ Three or four **muffin tins** partly filled with **tempera paints** in different colors

✔ For back-up: **cotton swabs or cotton balls**

Beginning

Gather around the long piece of paper and remind children that they've just gotten new toothbrushes. Ask two of the children in the group to pass out the old toothbrushes, one to each child. Set out the paint tins and explain, "Today, instead of using your old brushes to clean your teeth, you can use them to paint designs on the paper."

Middle

Watch the ways children paint, being supportive of the different ways children use the paint. Recognize each child's individual effort. For example, some children will paint with just one color, others may combine colors to make new colors, and still others may use the colors to represent a person or a thing. Choose strategies for supporting children based on their actions (for example, to the child painting with one color, "You're only using blue today"; to

Some children use only one color, while some mix colors.

Moving small-group time to the floor offers extra space to paint and a fresh perspective.

the child mixing colors, "I don't see orange paint in the cups anymore—how did that happen?"; or to the child making a representation, "Your person has a big smile—that usually means to me that he's happy"). To gain insights into children's developing social abilities, notice the ways children interact with one another when they want to switch colors or borrow ideas. Help children who seem to be having difficulty expressing their needs ("You're pointing to the blue paint, Mary—you could say 'Pass the blue paint'").

End

Tell children that the group is about to end and that you will be bringing two tubs of soapy water to the table for cleanup. Ask children to wash their brushes and put them away in a labeled container that will be stored in the art area. Observe whether anyone notices and comments on the fact that the toothbrushes now used as painting tools are stored differently from those used for brushing teeth. Engage children in helping you find a suitable drying spot for their paintings.

Children's cooperative abilities develop as they trade or borrow paints.

Follow-up

1. Hang the mural in the bathroom or other area where children brush their teeth in your center.
2. Store the old toothbrushes in the art area.

Sun Reflections

Originating ideas

Lately, some children have shown a fascination with the way the sunshine moves around the classroom at different times during the school day. Children have been making plans to "play in the sun patch" on the floor near the toy area, and they have been putting their hands on the walls to see if they can make shadows when the sun is shining on the walls.

Possible key experiences

Classification: *Using and describing something in several ways*

Space: *Experiencing and describing positions, directions, and distances in the play space, building, and neighborhood*

Time: *Experiencing and describing different rates of movement*

Materials

✔ A variety of **hand-held mirrors,** one per child

✔ For back-up: **hand puppets** (for making shadows)

Beginning

After meeting in your regular small-group meeting space, move to a space large enough for children to move freely. Tell them you've brought something to small-group time that is small enough to hold in their hands and that will make the sunlight dance and jump on the walls, ceiling, and floor. Ask them to walk around the room (don't pass out the mirrors yet) and see if they can find a patch of "magic sunlight." Then use one of the mirrors to make a spot of reflected sunlight somewhere else and wait for children to find the spot and describe where it is. Notice if they use simple sentences ("There it is!") or if they use details in their descriptions ("It's high up above the toy shelf!")

Middle

Give each child a mirror and watch as they make moving and stationary sun patches on the walls, ceiling, and floor. See if they play alone at this, or if they make games with one another. When you enter their play, take on a role that connects to their choices (for example, sit next to a child who is playing alone and imitate his or her actions; join a small group of children who are playing a game of putting their hands and feet in one another's sun patches by putting your own hand on a sun patch). Encourage children to move to different parts of the room and to make the reflections from different positions (sitting on the floor, standing on a chair, lying on the floor). Comment on how quickly or slowly their sun patches are moving, and comment on their body positions.

End

To bring the small-group time to an end and start cleanup, ask children to freeze so their sun patches will stay in one place. Touch a child's sun patch, and ask that child to put his or her mirror away. Then have that child touch another child's patch, and have that child put his or her mirror away. Repeat this process until all the mirrors are put away.

Follow-up

1. Add the mirrors to the toy area.

2. At large-group time play various musical selections and ask children to pretend they are the patches of light dancing on the floor. Alternate between slow and fast selections. For example, "Ersko Kolo" from the High/Scope Press recording *Rhythmically Moving 4* has slow and fast sections that alternate.

3. Hang prisms in various locations in the classroom so children can experience the sunlight reflecting in a new way.

Animal Care

When teachers observed Demetrius and other chilren engrossed in pretending about dogs, they decided to bring in a real dog for small-group time.

Recently, children have been pretending to be cats and dogs. In these pretending games, some children crawl around on the floor while others guide them on "leashes" (strings tied to children's wrists) and brush their "fur." To build on this interest, the teachers have planned for Alex, a staff member's patient dog, to come in for a visit.

Possible key experiences

Creative representation: *Recognizing objects by sight, sound, touch, taste, and smell*

Initiative and social relations: *Being sensitive to the feelings, interests, and needs of others*

Materials

✔ Alex, the **dog**

✔ **Pet accessories: Combs, brushes, dog treats, a bowl of water, hair barrettes or ribbons, Alex's leash**

✔ Materials for making leashes: **string or ribbon, scissors, tape**

✔ For back-up: **stuffed animals**

Beginning

Gather around Alex in a large space. Give children time to touch, smell, and follow Alex, and to listen to any sounds he makes. Provide any safety information that is necessary, for example, "Alex doesn't like it when you touch his tail." After a while bring out the additional materials listed above.

Middle

Position yourself with the dog at the children's level, and listen for the children's comments about his size, the softness of his fur, the smell of his breath, and the sounds or motions that

Experiences with a real dog provide a concrete understanding of what dogs do and how people care for them.

he makes to "talk" to them. Watch the ways that children touch or approach the dog: Do they sit right next to him? Do they watch him, but from a distance? Are they completely uninterested in the dog, instead playing with the pet accessories? Acknowledge their comments ("He *does* pant with his tongue when he's thirsty"), give them additional information ("He's wagging his tail when you brush him there—that means he likes it"), and ask them questions ("He's stretching his legs—what do you think that means?").

The small-group ends with a brief dog walk. Children take turns holding the leash.

End

Give Alex one of his treats. Tell children that they have about one more minute to be with Alex indoors and that when he finishes his treat it will be time to put on his leash and take him for a walk. When a minute has passed, put away the grooming items and take Alex outdoors for a short walk. If necessary, work together with children to decide on a turn-taking system for holding the leash.

Follow-up

1. Make wrist leashes for children. At planning time, have each child use a leash as they walk a partner to the area in which they plan to work. Then have the child describe his or her plans.

2. Visit a pet store or kennel.

3. Add empty dog food containers and water bowls to the house area.

Popsicle Stick Play

Originating ideas

After finishing a snack of frozen juice bars on sticks, several of the children began playing with their sticks instead of throwing them away immediately. They combined the sticks to make a variety of shapes, letters, and designs on the table.

Possible key experiences

Language and literacy: *Writing in various ways—drawing, scribbling, letterlike forms, invented spelling, conventional forms*

Classification: *Sorting and matching*

Materials

✔ Wooden and plastic **Popsicle and juice bar sticks**

✔ **Twigs and small branches** from outdoors

✔ For back-up: **Cardboard pieces** and **glue**

Beginning

Bring the sticks and twigs to the table, and arrange some of them to make a shape, such as a rectangle. Say to the children "Yesterday I saw children making designs and letters with their Popsicle sticks from snack. Today I brought enough sticks for everyone to make their own designs by themselves or working with others. See what you can do with them." Give each child an individual basket filled with sticks, twigs, and Popsicle sticks.

Middle

Begin playing with the materials yourself, shaping and reshaping your arrangements of sticks. As you work, watch to see what the children are doing. Some children may use the sticks to make letters, shapes, or other recognizable symbols or patterns. Some may use their sticks to make props for dramatic play (for example, brandishing them as if they were swords, sucking them as if the frozen bars were still there, piling them up to make a campfire). Others may line up and sort the sticks so that they have a pile of popsicle sticks next to a pile of twigs and small branches. Join in children's play by following the play direction chosen by the child ("I see you've made a big stack of sticks. Would you like one of mine to add to your pile?" or "You made a **T** with your stick—Did you know that is the first letter in Tamara's name?").

End

Bring two containers to the table, one for leftover Popsicle sticks, the other, for twigs and branches. Watch as children put away their remaining materials to observe the ways they sort and match the items.

Follow-up

1. Add Popsicle sticks, twigs, and branches to the art area and to the sand and water table.

2. Bring a stick to the next day's planning time, and ask children to use it to point to an area they will work in or to touch an object they will use in their work time plans.

3. When classroom supplies of twigs and small branches need to be replenished, ask interested children to come along with you as you take a walk around your outside space to collect additional materials.

This activity was planned after a snack of Popsicles, because teachers had observed children playing with the sticks after they were finished eating.

Tissue Paper Color Mixing

Originating ideas

During a recent work time, children experimented with and commented on the ways the paint colors they were mixing at the easel blended together to make new colors.

Possible key experiences

Language and literacy: *Describing objects, events, and relations*

Initiative and social relations: *Making and expressing choices, plans, and decisions*

Seriation: *Comparing attributes (longer/shorter, bigger/smaller)*

Number: *Arranging two sets of objects in one-to-one correspondence*

Materials

✔ **Tissue paper squares** in a variety of colors in individual **containers** for each child

✔ A large **clear plastic bowl** or other **container** filled with **water**

✔ **Water** in **clear plastic cups** (Have several pitchers of water ready to refill cups as needed.)

✔ **Stirrers** (wooden coffee stirrers or plastic spoons)

✔ **Newspaper**

✔ **Smocks**

✔ **Paper towels**

✔ For back-up: **teddy bear counters**

Beginning

Have children put on smocks and help you cover the table with newspaper. Set the large plastic container of water in the center of the table, and give each child a tissue paper scrap. Encourage children to take turns putting their scrap in the water and stirring it around to see the color of the water change. Listen to the ways they describe the changes in the water and the texture of the paper.

Middle

Give each child an individual water container, paper scraps, and stirrers. As children experiment with the materials, listen for any comparison language they use. Challenge their thinking by repeating their discoveries and posing new questions ("First the water was light, but now it's getting darker"; "It did get darker when you added green—Do you think it will get lighter if you add yellow?"). Notice which children discard their colored water so that they can refill their cup with clean water. Some children may not use the water but will instead lay their paper scraps side by side or on top of one another on the floor or on the table. Talk to those children about their spatial arrangements or about the color changes that result when they lay paper scraps on top of one another. Some children make take their paper scraps out of the water to dry; if you see children doing this, provide paper towels to use as a drying surface and as another choice for children who enjoy watching the colors bleed from one surface to another.

End

Put children's names on their paper towel, if they used one. With children, empty out the water containers and throw newspaper and wet paper scraps into the trash can. As children are cleaning up, talk about the experience they've just had.

Follow-up

1. Add colored tissue paper scraps to the art or sand and water table areas.
2. Announce the addition of these materials on the message board by fastening several overlapping scraps to the board, creating a color-mixing effect. Next to the scraps, draw an arrow pointing to the symbol for the interest area where the paper scraps will be stored.

Ice Cube Hockey

Originating ideas

At outside time for the past several days, Audie, James, Kayla, and Julia have been gathering sticks and stones and using them to "play hockey." In this game they hit the stones with the sticks and then run after the stones.

Possible key experiences

Initiative and social relations: *Creating and experiencing collaborative play*

Movement: *Describing movement*

Classification: *Exploring and describing similarities, differences, and the attributes of things*

Time: *Experiencing and describing rates of movement*

Materials

- ✔ A rectangular **"hockey rink,"** made by enclosing a large section of a **tiled floor** (or a large **table top)** with **long unit blocks**
- ✔ **Popsicle sticks** or **small sticks** gathered from outdoors
- ✔ Several **bowls** of **ice cubes**
- ✔ Cloth **rags** or **paper towels**
- ✔ For back-up: **stones** and **toilet paper rolls**

Beginning

Meet with children sitting on the floor around the outside edges of the hockey rink, and give each child a Popsicle stick or small stick from outdoors. Take one ice cube out of the bowl, and use your stick to pass it across the rink to a child sitting opposite you. As you do this, describe your actions and the speed at which the ice cube is moving.

Middle

Pass out a cube to each child. Hit the cubes back and forth with the children, playing by the rules they suggest ("Hit it right back to me" or "Try to keep it away from me"). Listen for the ways children describe what they are doing and how the cubes are moving ("This one's going to

Sarah, and it's going to be fast"). Build on their descriptions ("I'm sending one to Sara, too, but this one's even faster"). Notice which children talk about the rules of real hockey games or try to establish rules, and the ways the other children in the group respond to their suggestions.

End

To bring the small-group time to an end, put two plastic bowls in the middle of the hockey rink. Ask children to try and toss their cubes into a bowl. Give each child a cloth rag or paper towel. With children, wipe any blocks that are wet, put the blocks away, and wipe up any water left on the floor.

Follow-up

1. Repeat this activity, keeping the ice cubes but using a different playing surface: a large piece of paper taped to the floor with dollops of wet paint scattered across it or with dry tempera paint sprinkled on it. Watch as the ice cubes make paint designs on the paper.

2. Add plastic hockey sticks and pucks to your outdoor toy collection.

Making Bookmarks

Originating ideas

One of the teachers has been reading a long book to the children one chapter at a time. Recently, one of the children noticed that the teacher, upon finishing a chapter, used a bookmark to mark her place. The child immediately asked the teacher what she was doing. When the teacher explained what a bookmark was and why she used it, other children asked to see the bookmark.

Possible key experiences

Creative representation: *Drawing and painting*

Language and literacy: *Reading in various ways—reading storybooks, signs and symbols, one's own writing*

Classification: *Sorting and matching*

Time: *Anticipating, remembering, and describing sequences of events*

Materials

✔ Bookmark-sized **paper strips** made from heavy oak tag paper

✔ **Markers, stickers, stamps,** and **ink pads** for decorating the paper

✔ A **book** with short chapters or poems, such as Shel Silverstein's *Attic*

✔ For back-up: a collection of other **storybooks** (These do not have to be "chapter books.")

Beginning

Read a chapter or poem from the long book you have selected. When you finish, ask children if they remember what people can do to mark their place in a book, so that the next time they read they'll know where to start again. Accept the children's definitions, descriptive words, and other ideas about bookmarks, and repeat their language back to them. When you finish reading, pass out the materials, and tell children they may use them to make their own bookmarks to leave at school or to use with books they might read at home.

Middle

Comment on the stickers and stamps the children are selecting for their bookmarks and listen to the ways they describe their choices. Describe your choices as you make your own bookmark ("I think I'll make this one with all animals, then use it with books about animal stories"). Observe which children simply decorate their paper strips, and which children demonstrate their understanding of a bookmark's purpose by taking their finished paper strips and using them with books from the book area or the back-up books you have collected. Be equally supportive of either choice—don't push the children to recognize their paper strips as bookmarks.

End

Put children's symbols and names on their bookmarks, encouraging those children who are capable of it to write their own symbols or names. Suggest to children that they may leave their bookmarks at school to use with the books in the book area, or take them home to use with their own family's books.

Follow-up

1. Add children's bookmarks to the book area.

2. Get out the leftover paper strips, and write an area symbol on the top of each one. At planning time, hold the strips like playing cards, fanned out in your hand, with the symbols facing you so the children can't see them. (You may prefer to ask a child to hold the strips.) Have one child at a time choose from the strips in your hand, and then ask the child to tell you whether his or her plans include the area represented on that paper strip.

This child illustrated a number sequence with the triangle groupings on her bookmark. She demonstrated her classification skills by choosing a different color for each group of triangles.

Snow or Sand Painting

Originating ideas

The other day, as one of the children was leaving school with a wet painting, it dropped to the ground in the snow. When he retrieved it, he noticed that the wet paint had made a mark on the snow. This caused a flutter of excitement among the children, who thought it was funny to see colored snow. The next morning at greeting time, two children mentioned that the color was still there.

Possible key experiences

Creative representation: *Drawing and painting*

Classification: *Using and describing something in several ways*

Seriation: *Comparing attributes (longer/shorter, bigger/smaller)*

Materials

✔ A **sand and water table** or several **large plastic tubs** filled with **snow** (Note: in warmer climates, substitute **sand** for snow.)

✔ **Paintbrushes** of different widths and **tempera paint** in various colors

✔ **Smocks**

✔ For back-up: **paper**

Beginning

Remind children about the colored snow outside, and tell them you have brought the snow inside so they can paint on something other than paper. Then gather around the table or plastic containers and set paint and brushes nearby so children can begin painting.

Middle

Since this is such a new material, don't be surprised if children paint without interacting with one another or talking. Observe children carefully. Do they use more than one size of paint-brush? One or several colors? As children work quietly, sit near them, imitating their actions. Base your comments about your own work on what you see children doing ("This brush made a much thicker mark than the ones in Alana's snow"). If a child doesn't react to your comment, stop talking and paint quietly next to him or her. If children engage in conversation, listen for the topics they choose: Do they compare paintbrush sizes or talk about the differences between painting on snow and on paper? Do they share personal stories as they work with the snow and paint? Make sure that any comments you make reflect the interests children express.

End

Take some of the painted snow and scoop it into a bowl to store in the freezer. Leave the remainder of the snow out so children can compare what happens to the two batches of snow. With children, cover the paint containers, clean the paintbrushes, and put away the materials.

Follow-up

1. Bring to greeting time a tub of frozen painted snow and the tub of water from the snow that was allowed to melt. Encourage children to compare what happened to the snow and the colors.
2. Refill the table with fresh snow, and see if children make plans to paint again.

Bagel Creations

Originating ideas

At work time, some of the children have been making models of people by combining large balls of Play-Doh with smaller balls (which they call the eyes and the noses). To complete their Play-Doh people, the children insert toothpicks to represent arms, legs, and hair.

Possible key experiences

Creative representation: *Making models out of clay, blocks, and other materials*
Initiative and social relations: *Making and expressing choices, plans, and decisions*
Classification: *Using and describing something in several ways*

Materials

✔ **Bagel halves,** one per child
✔ Several containers of **cream cheese,** in assorted colors (Either buy the flavored kind, or color plain cream cheese with food coloring.)
✔ Plastic **knives or thin spatulas**
✔ **Bowls** filled with **sliced carrots, cherry tomatoes, sliced black olives, sliced green and red peppers, sliced cucumbers,** and **alfalfa sprouts**
✔ **Paper plates** and **napkins**
✔ For back-up: a pitcher of **water** and **cups**

Beginning

Tell children that you have noticed them decorating Play-Doh pieces at work time, so you've brought some new kinds of materials for them to decorate.

Middle

Pass out plates, bagels, spatulas or knives, and the cream cheese, reserving the other ingredients. As children spread their cream cheese, talk to them ("I notice you only used blue"), and see if they offer explanations of their choices ("It's because my sister likes blue"). While children are spreading cream cheese, place the bowls of vegetables and other trimmings on the table and tell children these are extra things for decorating their bagels. Note which children simply eat the food. Ask questions about how things taste or what the children like most, and listen for any descriptive language children use ("It's crunchy/salty/etc."). Watch for children who use the food to make a representation (a person, a monster). Converse with children about their work when invited, moving on to another child if a child ignores you.

End

Tell children that it will soon be time to eat their bagel creations. As children eat, make up a guessing game about the foods they like best on their creations ("I'm thinking your favorite food is crunchy and hard").

Follow-up

1. Bring bagel-shaped paper to the next planning time. Provide markers and ask children to decorate their papers with their work time plans; they could do this by getting an item they plan to work with and tracing around it or by drawing a picture of what they plan to do.

2. Serve bagel halves with peanut butter right after small-group time, encouraging children to make comparisons between the peanut butter and the cream cheese.

Washing and Hanging Clothes

Originating ideas

As part of their plan to "take care of the babies" at work time the other day, Cirra and Asia took the baby clothes from the house area to the sand and water table, dribbled liquid soap inside, and scrubbed the clothing by hand. Several other children changed their work time plans to join the two girls in washing baby clothes and other items; some of the children washed the dress-up clothes from the house area that they had put on for pretend play. When cleanup time started, there was a pile of wet clothing and nowhere to hang it to dry, since the drying rack used for children's artwork was already full of children's paintings.

Possible key experiences

Movement: *Moving in nonlocomotor ways (anchored movement—bending, twisting, rocking, swinging one's arms)*

Classification: *Sorting and matching*

Seriation: *Comparing attributes (longer/shorter, bigger/smaller)*

Number: *Arranging two sets of objects in one-to-one correspondence*

Time: *Experiencing and comparing time intervals*

Materials

- ✔ A **basket** of dirty **clothing** or **cloth pieces,** enough for each child to wash several articles during the course of the small-group time
- ✔ Large plastic **laundry tubs (or a sand and water table)** filled with **warm sudsy water** and, if possible, set up outdoors
- ✔ A **clothesline,** hanging in a partly sunny, partly shady area, if possible (Hang the line low enough so that children can reach it easily.)
- ✔ **Clothespins**
- ✔ For back-up: an assortment of **water toys** (for example, plastic containers, funnels)

Beginning

Gather outdoors around the filled laundry tubs or water table, and tell children that you have collected a pile of dirty clothes for washing, some that fit the baby dolls and some that fit the children. Remind them of the problem created at work time where there was a pile of wet clothes and no place to dry them. Point out the clothesline and set out the clothespins.

Middle

Scrub clothes alongside children. Listen and observe, noting whether their conversations or actions involve sorting the clothing into piles according to size or other attributes. If they do sort, play sorting games with them, asking them if they want your clothes added to their piles. As they hang up clothing, watch the way the children manipulate the clothespins for clues to their manual coordination and problem-solving skills. Help them make comparisons about the differences in temperature between the sunny spots and the shady spots on the clothesline, encouraging them to predict where clothing will dry sooner.

End

Work with the children to empty and dry out the wash tubs or the sand and water table. Put away unused clothespins in a container, and put any extra clothing that has not been washed back in the laundry basket.

Follow-up

1. At outside time or at the end of the day, take the children to check whether the clothes are dry. If they are, help the children take them off the line and put them into a laundry basket.
2. Add empty laundry soap boxes, a toy washing machine (either purchased or homemade), and a laundry basket to your house area.

Teachers planned a clothes-washing activity when they saw Cirra and Asia washing baby clothes in the water table.

Jewelry Design

Originating ideas

The mother of Janine, one of the children in the classroom, wears a variety of rings, bracelets, and necklaces that she changes often. Some of the children have noticed and made comments about this. Toward the end of one day's session (before Janine's mother had arrived to pick her up) one of the children said, "I wonder what kind of earrings she'll have on today."

Possible key experiences

Creative representation: *Making models out of clay, blocks, and other materials*

Initiative and social relations: *Solving problems encountered in play*

Space: *Changing the shape and arrangement of objects (wrapping, twisting, stretching, stacking, enclosing)*

Materials

✔ Brightly colored **pipe cleaners**

✔ **Fasteners: twist ties, rubber bands,** and **paper clips**

✔ **Ribbon**

✔ For back up: **paper** and **glue**

Beginning

Give each child his or her own container with the assortment of materials described above. Say "I noticed that you have enjoyed Janine's mother's jewelry when she comes to pick up Janine. Here are some things you can use to make your own jewelry."

Middle

Watch as children twist, tie, and manipulate the materials, observing for information on how children use the materials for representation. Notice which children simply explore the materials—arranging and reshaping the paper clips and pipe cleaners, for example—and which children use them to make specific objects or representations (jewelry, people, eyeglasses). Watch for problems (how to make a bracelet long enough to fit around a wrist, how to attach one clip to another so that the necklace doesn't slide off), and observe the ways children solve them. Observe how independent and persistent children are in solving problems that arise. Do they give up after one try? Ask for help? If so, offer a suggestion ("Did you try. . .?"). Do they successfully complete the task on their own? If so, acknowledge their accomplishment ("You worked hard to tie that twist tie around the paper clip"). Is there conversation while the children work, about Janine's mother, their own parents' jewelry, or about what they are doing? Make your own jewelry creation, and if children are willing, ask them to describe how they made something work so you can incorporate their ideas into your own design.

End

Beforehand, set aside space for a jewelry display. Lay out blank pieces of paper in the space, and label each sheet with a child's name and personal symbol. When small-group time is almost over, wave a pipe cleaner as if it were a magic wand, and tell children that when the magic wand touches each child it will be his or her turn to put away materials. As they take their turns, have children lay their creations on the papers identified with their personal symbols and names.

Follow-up

1. Add twist ties, pipe cleaners, rubber bands, paper clips, and ribbon to the art area.

2. Put costume jewelry in the house area.

3. Ask Janine's mother if it would be possible for her to sit at the planning table with the children and let them wear one of her pieces of jewelry while they tell others their plan.

4. Take photos of Janine's mother wearing her jewelry and of children wearing the jewelry they made, and hang up the photos near the dress-up clothes.

Monster Protectors

Originating ideas

Several times recently at greeting circle, children have asked adults to read *Abiyoyo* (as retold by Pete Seeger). Some children's work time plans have included acting out various parts of the story (the fast dancing, the monster disappearing).

Possible key experiences

Creative representation: *Making models out of clay, blocks, and other materials*

Language and literacy: *Having fun with language—listening to stories and poems, making up stories and rhymes*

Initiative and social relations: *Solving problems encountered in play*

Music: *Exploring and identifying sounds*

Materials

✔ Long **cardboard tubes or plastic wands**

✔ **Materials for decorating** the tubes—**glitter, glue, streamers, masking tape**

✔ **Sound-making materials—beans, pebbles, sand**

✔ The storybook ***Abiyoyo***

✔ For back-up: **markers**

Beginning

Bring the book to the small-group meeting place. Before opening it, encourage children to recall details of the story. Repeat their interpretations and ask questions or make comments connected to their ideas. For example, if a child says, "The monster chased the people" you could say "The monster chased the people. What did the people do?" When children stop talking, read the story along with them, again pausing for their comments and questions. When you are finished reading the story, bring the other materials to the table and say "I wonder if you could use these materials in some way to make a magic wand or a musical instrument to protect you from the monster." Give each child a tube or wand, and set out additional materials in the center of the table.

Middle

Work alongside children making your own magic wand or musical instrument. While you work, notice the complexity of children's chosen activities and interact accordingly. For example, "You're covering the tube with glue—would you like to add some glitter?" is one possible response for a child using only glue and a tube. On the other hand, to a child whose project is more complex (stuffing beans inside a tube and taping the ends to make a musical instrument), you might say "Give me a clue about what is inside, and I'll see if I can guess." Note whether any children refer back to the actual story line as they make their creations or simply become involved in the process of making something or exploring the materials. Listen for children discussing their work with others to see if they are remembering details from the book that they plan to use later ("I'm gonna take this outside and use it to chase away the monsters").

End

To facilitate the cleanup process, act out the part of the book where the wand makes things disappear and the music goes faster. Touch objects on the table with your wand "so they will disappear into their containers." Or make sounds with the instrument you or a child has made when you want cleanup to go faster.

Follow-up

1. Act out parts of the story at large-group time, using the child-made instruments "to make the monsters dance faster."

2. At the next planning time bring a decorated tube or plastic wand to the table, and place a set of area symbol cards face down on the table. Ask a child to pick a card, then shake the tube or wand and have children who plan to go to that area "disappear" from the table.

Tube and Cup Sculpture

Originating ideas

Lately, some children have been making collages with three-dimensional materials, such as noodles, cardboard tubes, and cotton balls. Teachers planned this activity to give all the children additional opportunities to make three-dimensional creations.

Watching Christopher figure out how to fasten three-dimensional materials to paper inspired teachers to plan the "Tube and Cup Sculpture" activity.

Possible key experiences

Seriation: *Comparing attributes (longer/shorter, bigger/smaller)*

Space: *Changing the shape and arrangement of objects (wrapping, twisting, stretching, stacking, enclosing)*

Materials

- ✔ Large flat, sturdy **pieces of cardboard,** one per child
- ✔ **Paper cups** of different heights
- ✔ Cardboard **toilet paper or paper towel tubes** cut into segments of various heights
- ✔ **Paper clips, glue, glue brushes,** and **tape**
- ✔ For back-up: **markers**

Beginning

Give each child a piece of cardboard. Say "Right now this is empty, but I'll put materials on the table you can use to fill it up."

Middle

Put the tubes, paper clips, glue, tape, and cups in the middle of the table. Observe the various ways children combine the materials. Some may attach the cups together using the paper clips, while others may brush glue on the cardboard and press the tubes and cups onto the glue. Still others may attach the cups to the tubes with the tape, spending the whole small-group time taping the pieces together to make one structure. However the children combine the materials, be available to comment on their work when an opportunity for conversation arises. Encourage them to describe what they are doing. For example, you might say something like this, "If I wanted my castle towers to look like yours, what should I do first?"

End

When small-group time is almost over, remind children to put their work on their cardboard pieces. Put away their creations (labeled with children's names and personal symbols) in a space where they can dry. Tell children you will be using them again the next day, with additional materials for decorating.

Follow-up

1. Bring the sculptures to the next day's small-group time and have children decorate them (see "Decorating Tube and Cup Sculptures," p. 168).
2. Tape two toilet paper rolls together and use them as pretend binoculars at planning or recall time. Ask children to use the binoculars to look at the materials or toys that they plan to use (or did use) at work time.
3. At large-group time, have children put each of their hands inside a paper cup. Then have them tap the ends of the cups together or tap the cups on different body parts, matching the beat of a musical selection. When the large-group time is over, have them clean up the cups by stacking them together cooperatively to make a tall sculpture.

Paper-Water Experimentation

Originating ideas

During the small-group time "Colored Water, Eyedroppers, and Suction-Cup Soap Dishes" (p. 90), children were fascinated by the cleanup process. As they wiped up the excess water with paper towels, they said things like "Look, it's moving," "The colors are running," "It makes the table dry if you rub it hard, but then the paper is wet." Teachers decided to expand on children's interest in the effects of water on paper by providing a variety of paper types for children to experiment with.

Possible key experiences

Classification: *Exploring and describing similarities, differences, and the attributes of things*
Seriation: *Comparing attributes (longer/shorter, bigger/smaller)*
Time: *Experiencing and describing different rates of movement*

Materials

✔ **Colored water**

✔ **Clear plastic cups**

✔ Plastic **eyedroppers**

✔ Samples of each of the following paper types for each child: **paper towels, construction paper, waxed paper, aluminum foil, plastic bubble wrap** (the kind used as filler in packaging) and **coffee filter paper**

✔ For back-up: **glue** and **cardboard pieces,** large enough to serve as the base for a collage

Beginning

Put two different types of paper (waxed paper and coffee filter paper) in the center of the table. Give each child an eyedropper and cup of colored water, and ask children to put a droplet of water on the two different types of paper. Say "See what you can notice about what happens when you drop the water onto the paper." Remember the comments children make, and repeat them so others can hear the reactions of their classmates. When they are finished describing their efforts, pass out the baskets with the individual papers for each child.

Middle

Notice how children experiment with the materials. Do they select one piece of paper and fill it with water before moving on to another piece? Do they dribble water on several pieces at a time, then talk about the differences? If children notice differences, challenge their thinking with comments like "I wonder what happens to the water when you drop it onto the waxed paper or the paper towel." Share children's excitement about their discoveries by matching their enthusiasm when they describe their efforts to you. Imitate their actions by using your materials in the same ways that fascinate them (for example, dribbling water droplets on waxed paper, then holding the paper so that the droplets dance around).

End

Ask children to suggest a way to sort their leftover paper pieces, based on their own observations of what happened to the water when it dripped on various types of paper. Accept their descriptions of the categories they are sorting by. For example, "The water goes in" could be one category and "The water stays on top" could be another. If children's ways of sorting show that they don't understand the suggested categories, don't correct them. Instead, simply note for future reference those children who do classify in this way, as well as those who sort by different criteria.

Follow-up

1. Add to the art area a bin of the paper samples from this activity.
2. Announce the new materials and their storage place on the message board, discussing the message at greeting time. See if any of the children in the previous day's small-group time describe how the materials were used.

Eggshell Pictures

Originating ideas

During a snack time in which children peeled hard-cooked eggs, they were especially interested in the shells. As the shells came off, they talked about their textures and shapes, and the fact that you couldn't eat them, just the egg inside.

Possible key experiences

Creative representation: *Recognizing objects by sight, sound, touch, taste, and smell*

Classification: *Exploring and describing similarities, differences, and the attributes of things*

Materials

✔ Whole and crushed **eggshells**—some colored, some left white—for each child (Different colors can be made by soaking the shells in a mixture of water and food coloring, then letting them dry on paper towels.)

✔ **Construction paper**

✔ **Glue** and **glue brushes or glue sticks**

✔ For back-up: **markers**

Beginning

Begin small-group time with a guessing game, in which you give children clues about a snack they ate recently. Describe some of the attributes of hard-boiled eggs (and the shells they peeled off them) until children guess what the small-group material of the day is.

Middle

Pass out the eggshells, paper, and glue. Tell children that since the shells are not for eating, you've brought them glue and paper to use in making pictures. Watch as children begin their work. Some will simply play with the shells, sorting them by colors or crushing the ones that were left intact. Listen to the ways they compare the colored and white shells and the words they use to describe the small pieces. Other children will glue shells on paper to make designs. To gain an understanding of the ways children classify objects, watch to see if children glue a mixture of eggshells in one big pile or keep the different-colored shells separate. Note whether any children label the work they are creating to indicate that it is intended as a representation of something.

End

Collect extra eggshells in a single container, stirring with each addition to make a collage of color. Ask children if they want to describe their work to others before storing it to dry.

Follow-up

1. Bring an eggshell picture to the next day's greeting time. Explain that there are some eggshells left to make new pictures with at work time, but not a lot, and that confetti and colored paper scraps have also been added to the art area as extra materials for making gluing pictures.

2. Ask parents to save eggshells from home, and designate a collection spot for them in the classroom. When you have collected enough, replenish the supply in the art area.

Fabric Banner Painting

Originating ideas

On a recent walk around the school neighborhood, children noticed some houses with hanging banners. They began a game of looking for houses with banners and comparing the painted pictures on the cloth. They spotted banners with a giant sunflower, a hummingbird pecking at a red flower, and a sun.

Possible key experiences

Creative representation: *Drawing and painting*

Language and literacy: *Talking with others about personally meaningful experiences*

Space: *Interpreting spatial relations in drawings, pictures, and photographs*

Materials

✔ **Pieces of thick cotton fabric,** in solid colors, one per child

✔ **Paintbrushes** and containers of **tempera paint**

✔ **Smocks**

✔ **Newspaper**

✔ **Photographs** taken of the neighborhood walk, including images of houses displaying banners, houses without banners, and children walking down the sidewalk

✔ For back-up: **markers**

Beginning

Pass around the pictures taken on the walk. Listen as children describe what they see in the photographs and expand on the details children recall by talking about things related to the pictures. Encourage children to think about spatial concepts by posing questions about spatial relations between objects in the photos ("Look at this house. Do you remember if we were closer to the mailbox on the house or the banner when we were standing on the sidewalk?"). Tell them you have materials for them to use at small-group time that will remind them of something they saw while on the walk. Give them clues and see if they can guess ("It was hanging near some people's houses, but not near others"; "You'll need a smock, and a large space to work"). After children have put on their smocks, spread newspaper on the painting surface and give each child a fabric piece. If necessary, help children find spaces large enough that they can work comfortably on their banners.

Middle

Comment about children's work as they are painting on the fabric. You might mention the designs they are making, their color choices, and the areas of the cloth they are covering. Do not expect children to make pictures on their banners like those on the banners they saw on their walk, but comment on the similarities if they do ("You made a round, yellow circle. That reminds me of the sun we saw yesterday on the banner at the blue house").

End

Since children showed an interest in comparing the pictures on the banners they saw on the walk, see if they would like to compare the pictures they created during the small-group activity. To encourage this, you might walk around together to take a look at everybody's banners. Set out some soapy water tubs to soak the wet paintbrushes, and ask children to hang up their banners to dry in an appropriate spot.

Follow-up

1. Add sticks or dowels to the art area, in case children want to make their paintings into flags or banners.
2. Give each child a pre-made or purchased banner to move with at large-group time.

Balloon Toss

At outside time children have been throwing and catching balls of different weights and commenting on their lightness and heaviness.

Possible key experiences

Movement: *Moving in nonlocomotor ways (anchored movement—bending, twisting, rocking, swinging one's arms), moving in locomotor ways (nonanchored movement—running, jumping, hopping, skipping, marching, climbing),* and *describing movement*

Classification: *Exploring and describing similarities, differences, and the attributes of things* and *using and describing something in several ways*

Seriation: *Comparing attributes (longer/shorter, bigger/smaller)*

Materials

✔ An assortment of **balloons filled with water, air, or helium** (Choose balloons of different sizes, shapes, and colors. Make sure you have enough balloons to give children the options of playing alone or in small groupings. Be sure to secure any helium balloons with weighted strings so they don't fly away.)

✔ Extra **clothing** (for use if children get wet)

✔ For back-up: soft **foam bats** or **cardboard tubes**

Beginning

Gather indoors at the regular small-group table, and show children samples of the different kinds of balloons. Say something like "Today I brought balloons for you to play with but they're not all the same." Listen for the words children use to describe any similarities or differences. Repeat their words, adding details when appropriate ("These are both orange, and one is lighter than the other"). When they are finished talking, explain that you will be playing with these balloons in a larger space outdoors (or in a gym).

Middle

Go outdoors (or to the gym) where the rest of the balloons are located. Catch balloons when children throw them to you, stop and listen when they make discoveries they want to share with you ("It broke, and the water spilled out when I threw it on the sidewalk"), and make up your own games that children may choose to incorporate in their play (for example, stand on the weight attached to the helium balloon and use it as a punching bag). Find ways to comment on how children are moving ("You stretched up really high to catch that balloon while your feet were running").

End

Play a cleanup game of tossing the balloons back into their original containers, using comparison language: "First, put the heaviest balloons away." Ask children for their suggestions on which balloons to put away next.

Follow-up

Leave the balloons indoors overnight and bring them back outdoors the next day. Watch to see if children make comments about the ways the balloons have changed overnight or stayed the same. Do this over a series of days so children can watch the changes as the balloons lose air, leak water, or break.

Plastic Worms and Lizards

Originating ideas

At outside time Alex, Kayla, and Kevin have been digging for worms, then bringing them to the adults and the other children and pointing out how "slimy" and "wiggly" they are.

Possible key experiences

Creative representation: *Relating models, pictures, and photographs to real places and things*
Language and literacy: *Describing objects, events, and relations*
Classification: *Sorting and matching*

Materials

✔ An assortment of **plastic worms and lizards** (available in the fishing departments of discount stores) in **individual paper bags** for each child
✔ Clear **plastic cups**
✔ Pitchers **of water**
✔ **Paper towels**
✔ For back-up: **books featuring worms, lizards, or other "creepy-crawly" animals**

Beginning

Give a bag of worms and lizards to each child. Say something like "Dig inside the bag, and you'll find something that you would usually find outside."

Middle

Watch as children examine the contents of their bags. Note whether they compare the plastic worms to the real ones they have seen; listen for descriptive language that focuses on similarities or differences ("These wriggle like real worms but the real ones are brown"). Notice if children spontaneously sort the lizards from the worms or make groupings based on other attributes. Inquire about their groupings by describing what you see ("You put all the orange worms and lizards in the same pile"), then pausing to see if children offer additional explanations. If some children are losing interest, bring out the cups and pitchers of water, and note whether children use the worms and lizards with the water. Some may enjoy watching the plastic animals float, while others may stuff the animals into the containers and push them to the bottom. Add the paper towels so children can see the imprint the wet worm or lizard makes on the paper towel.

End

Ask children to line up the wet worms and lizards on paper towels to dry, and to put the dry ones in a basket in the middle of the table. Wonder aloud about where you might store the plastic worms and lizards in the classroom. Acknowledge children's responses.

Follow-up

1. Put worms and lizards in places suggested by children (for example, some may be stored in the toy area, others may be stored near the sand and water table).

2. Hide plastic worms and lizards in the sand at the sand table. At planning time, gather around the sand table. Give a shovel to the child whose turn it is to share work time plans. After the child digs up a worm or a lizard, he or she shares his or her plans with the others.

3. Make a snack (pudding, Jell-O) in which you have hidden gummy-candy worms.

Coloring Wood Chips

Originating ideas

At outside time recently, the teachers noticed children taking wood chips from the beds around the playground and making patterns with them on the sidewalk. While stones had already been added to the classroom for children to use in making arrangements and patterns (see "Stone Drawings," p. 164), teachers have noticed that children have been using the stones less and less often at work time.

Possible key experiences

Creative representation: *Drawing and painting*

Classification: *Exploring and describing similarities, differences, and the attributes of things.*

Materials

✔ Baskets of **wood chips** for each child

✔ **Markers**

✔ **Paint** and **paintbrushes**

✔ **Smocks**

✔ Three sample **wood chips,** one plain, another painted a bright color, and another marked or colored with a marker

✔ For back-up: **glue** and **heavy cardboard pieces**

Beginning

Bring the three sample wood chips to the table. Lay them out, and ask children to comment on their similarities and differences, questioning them about their explanations ("What makes you think the color is made with markers, and not paint?"). When the discussion is over, tell them you have a basket of wood chips for each of them and some materials they can use for making colors on them. Ask children to suggest places where they might arrange their wood chips if they do not want to color them.

Middle

Give children uninterrupted time to paint or color with markers. If you notice that they have stopped to examine their work, then comment on some aspect of it (the color blend, the texture of the paint on certain parts of the wood chips). Sit close to those children who decide to make arrangements with their chips, and imitate their actions. Again, occasionally talk about your own work ("I can line up these wood chips so they make a straight line"), or comment on the children's work when invited.

End

As part of the cleanup process, separate wet chips, chips colored with markers, and plain chips. Note which children, if any, sort along with you and which children collect all the chips together.

Follow-up

1. If you have a container of stones in the art area for making stone arrangements (see "Stone Drawings," p. 164), replace it with a container of colored and plain wood chips. Announce this change on the next day's message board.
2. Bring a container of wood chips to planning or recall time. Ask children to take out a wood chip for each of the areas they plan to work in (or did work in) that day.

II

Small-Group-Time Plans Originating From

New Materials

Finger Printing With Sponges

Originating ideas

At work time, instead of using paper and easels, children have been using paint and paint-brushes to paint on their hands and fingers. Once their hands are covered with paint, they use them to make imprints on paper.

Possible key experiences

Creative representation: *Relating models, pictures, and photographs to real places and things*
Classification: *Distinguishing between "some" and "all"*

Materials

✔ Thin **sponges** on **paper plates** and sheets of **construction paper** for each child
✔ **Tempera paint**
✔ **Newspaper** to cover the painting surface
✔ For back-up: an **assortment of objects** that children can use for paint-printing (**hair curlers, corks, plastic forks,** and **Bristle Blocks**)

Beginning

With children's help, spread news-paper over the painting surface. Give each child a piece of construc-tion paper and a paper plate with a sponge on it. As children watch, or with their help, pour tempera paint directly onto children's sponges. Wait a moment and watch as the paint soaks in. Say "I wonder what would happen if you touched the sponge with your finger and then put your finger on the paper?"

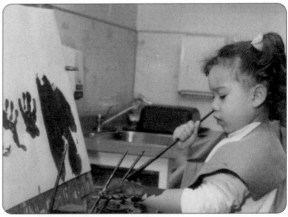

This child's interest in painting on her hands, then printing with them on the easel paper, gave teachers the idea for a small-group time that encouraged similar hand-printing activities.

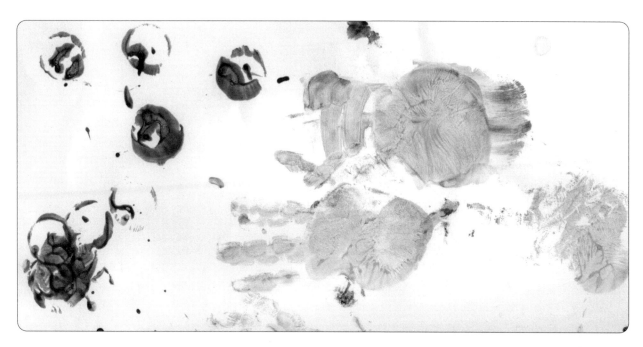

A child's artwork shows both handprints and prints made by dipping a small wheel in paint.

Middle

Listen to and support children's observations and activities. If they mention the way the paint is soaking into the sponge, crouch and look closely at the sponge. If they've made fingerprints on their piece of paper and they notice the swirls in them, look at the fingerprints with them. Then show them your own fingerprint swirls as a way of highlighting similarities and differences. To encourage them to try printing with other body parts, such as elbows, knees, or the palm of a hand, make up a guessing game with them ("I notice your fingerprint is on the paper. If you close your eyes, I'll put a different body part on my paper. See if you can guess which one").

End

Write children's names and symbols on their paintings and ask them to store them in your drying area. Have soapy sponges available for cleaning the tables, and encourage children to help wipe tables and stuff the paint-soaked newspaper into the trash can. Help children clean themselves, making yourself available as they wash themselves off to talk about the changes they notice in their hands and the colors of the water.

Follow-up

1. Repeat this activity outdoors on a warm day. This time, encourage children to paint and print with their bare feet.
2. Store three paper plate/sponge sets in a plastic bin in the art area to provide another painting option at work time.

Autumn Tree Treasures

Originating ideas

On a recent neighborhood walk children collected acorns, acorn tops, and horse chestnuts, with and without their cases. Before adding these materials to the classroom areas, teaching staff decided to use them in a small-group setting.

Possible key experiences

Creative representation: *Recognizing objects by sight, sound, touch, taste, and smell*

Classification: *Exploring and describing similarities, differences, and the attributes of things* and *sorting and matching*

Number: *Arranging two sets of objects in one-to-one correspondence*

After frolicking in leaves on a classroom outing, children began collecting acorns and horse chestnuts, which teachers used in a follow-up small-group time.

Materials

✔ A **shoe box** filled with **acorns and horse chestnuts** (Leave the top on and cut out a hole in the side big enough for a child's hand to fit through.)

✔ A small **bag** filled with **acorns and horse chestnuts** for each child (Make sure that some of the acorns still have their tops attached and that some of the horse chestnuts are still inside their cases.)

✔ **Magnifying glasses**

✔ For back-up: **teddy bear counters**

Beginning

Set out the specially prepared shoe box. Have children in your group take turns putting their hands inside to feel the contents, then describe and make guesses about what is inside. To encourage them to elaborate on their descriptions, make up funny responses based on what they say ("Hmmmm, something hard, it must be an elephant toenail").

Middle

Put the small bags and magnifying glasses on the table, and ask each child to take a bag. To gain information on how children sort materials and describe attributes, watch as children explore the materials inside. Observe to see whether any children put acorn tops on acorns, rub the textured surface of the horse chestnuts, or notice the dark or light markings on either material. Listen to the words they use to describe the spiny casings of the chestnuts ("It's prickly," "...like a porcupine," "Ouch!"). Notice if any children use the materials in role-play situations. Some children may arrange the chestnuts and/or acorns in lines, seriating them by size and labeling them as different family members based on their sizes. If they do, bring your own set of materials to where they are playing, and take on a role that relates to their play (for example, pretend to knock, using your acorn as a pretend person who says, "Hello, visitor for Grandma—Are you home, Grandma?").

End

As the group comes to an end, put two big containers in the middle of the table. Have children sort the acorns and chestnuts into the two containers. If children mix the two together when cleaning up, make a mental note, but resist the urge to correct them.

Follow-up

1. On the day after the small-group activity, bring a chestnut and an acorn to planning time. Ask the child who is planning to choose one of them to roll into the area he or she plans to work in. Then have the child describe his or her plans.

2. Add separate containers of chestnuts and acorns to the house area, toy area, or sand and water table.

Pomegranate-Apple Explorations

Originating ideas

At snack time several days before this small-group activity, lots of conversation was generated when kiwi fruit was served. Besides talking about the texture and color of the fruit, children also made comparisons between kiwis and other round fruits like apples and plums, even though apples and plums were not served that day. The teaching team decided to capitalize on children's interest in fruits by bringing in two fruits to explore, one familiar (apples) and one (pomegranates) that would probably be new to most children.

Possible key experiences

Creative representation: *Recognizing objects by sight, sound, touch, taste, and smell*
Classification: *Exploring and describing similarities, differences, and the attributes of things*
Seriation: *Comparing attributes (longer/shorter, bigger/smaller)*
Number: *Counting objects*

Materials

✔ **Red apples,** one per child
✔ **Pomegranates,** one for every four children to share
✔ **Plastic knives** for each child, and a **paring knife** for the adult
✔ **White butcher paper** to cover table
✔ **Paper towels** and **smocks**
✔ For back-up: **Magnifying glasses**

Beginning

Spread the butcher paper on the table with the help of the children. After the children have put on their smocks, roll an apple to each child, and put the pomegranates in the middle of the table. Wait for children's comments and questions.

Middle

After the children have explored the whole fruits, cut the fruits open with the paring knife so the insides are visible. Before you cut open each fruit, ask children to predict what they might see, then wait for their descriptions. Give each child apple and pomegranate sections to explore, using their plastic knives. Help children focus on the similarities of the apples and pomegranates (they are both red outside and rounded, they both have seeds inside, they both have white inside). Then encourage children to look for the differences (the pomegranate seeds are bright red and taste juicy and tart, while the apple seeds are dark and people don't usually eat them; the skin on a pomegranate is thick; an apple's skin is thin). To encourage children to comment on the differences, pose questions ("Do you think the red seeds will taste the same as the black seeds?") or make playful suggestions ("Hold out your hands and I'll put some skin in them—see if you can guess just by touching whether it is apple or pomegranate skin"). Notice whether any children line up the seeds from either or both fruits to count them or point out the way the pomegranate juice stains the white butcher paper.

End

With the children, collect leftover apple and pomegranate sections in a bucket in the middle of the table for disposal in your compost bin. Leave the paper out to encourage discussion about the red designs the juices made. Have children wash and dry their hands.

Follow-up

1. If possible, leave the butcher paper on the table for another day as a reminder of the activity.
2. At the next day's planning or recall time, have children place a pomegranate or apple seed on the area symbol card corresponding to their work time plans or experiences.

Exploring Packing Peanuts

Originating ideas

Children helped unpack boxes of art supplies delivered to the school and discovered packing peanuts inside. They were fascinated with them and began manipulating them in all kinds of ways, ignoring the art supplies inside.

Possible key experiences

Classification: *Exploring and describing similarities, differences, and the attributes of things* and *using and describing something in several ways*

Materials

✔ **Packing peanuts,** preferably **biodegradable** (the kind that dissolves in water), for each child

✔ **Clear, plastic containers filled with water,** one for each child

✔ Long **spoons, dowels,** or **coffee stirrers,** one for each child

✔ A **pitcher of water**

✔ Extra **plastic containers**

✔ For back-up: colored **construction paper** and **glue bottles**

Beginning

Give each children some packing peanuts and a plastic container filled with water. Say "I wonder what would happen if you combined the peanuts with the water." As they begin to explore the materials, add the stirring implements.

Middle

Listen to, watch, and support children as they experiment in various ways with the packing peanuts. Some children may choose not to combine the peanuts with the water, instead preferring to break the peanuts into tiny pieces and listen for the snapping sound this creates. Join in children's play and describe the results ("I heard a popping sound, then looked over and saw lots and lots of smaller pieces"). Other children may stuff their containers of water with peanuts and study the results. Encourage this curiosity by posing open-ended questions ("Some of your peanuts are on the bottom of the container, while others float on the top.

I wonder why?"). Still others may stir and then dump their mixture of peanuts and water onto the table, playing with and commenting on the texture of the newly made material.

End

Ask children to help you gather any dry packing peanuts that are left over and to find an appropriate storage place for them. Then pour some of the results of children's experiments into a large plastic tub, setting aside some of the mixtures in separate containers for storage overnight. With the children, stack the plastic storage containers and wipe up any spills. Check the remaining mixtures the next day to see if there are any changes in the materials, then discard them.

Follow-up

1. Bring the leftover peanut-water mixtures to greeting time for children to examine and discuss.

2. Add packing peanuts to the art area or sand and water table.

3. Soak ordinary, non-biodegradable, packing peanuts in water colored with food coloring. Allow peanuts to dry, then add the colored peanuts to your classroom interest areas.

Sometimes, to hold children's interest at small-group time, you may have to put out the back-up materials you've reserved. Here, a child uses glue and construction paper with the packing peanuts that were originally offered for the small-group activity.

Chopsticks, Tongs, and Tweezers

Originating ideas

At cleanup time the teachers observed two children making a game of putting away materials with a pair of tongs from the house area. They decided to add similar materials to the classroom that children could use for picking up various-sized objects. This activity is designed around these new materials.

Observing the ways children use materials when left to their own devices will help you think of ideas for small-group experiences. Here, Stone and Reid have made up a cleanup game using tongs from the house area. This gave teachers the idea for a small-group time planned around tongs and other grasping tools.

Possible key experiences

Initiative and social relations:
Solving problems encountered in play

Movement: *Moving with objects*

Classification: *Distinguishing between "some" and "all"*

Materials

✔ **Chopsticks, tongs,** and **tweezers** (Provide enough so that each child has at least one tool, with a few extras so children can experiment with more than one.)

✔ A **collection of small- and medium-sized objects** in different sizes and shapes (**inch-cube blocks, twisted pipe cleaners, wooden spoons, markers, paper clips, clothespins, small stuffed toys,** and **soft foam balls**)

✔ **Bowls**

✔ For back-up: **baskets** or other **large containers**

Beginning

Hide a pair of tongs in a paper bag and ask one of the children to guess what is inside. Older preschool-aged children may be able to give clues to the others at the table so they can guess. After pulling the tongs out of the bag, put an inch-cube block on the table and ask one of the children to pick up the block with the tongs and put it in the bowl. Then put a pair of tweezers on the table, and ask the children if anyone can pick up the block with the tweezers. Tell the children that for small-group time you brought some different materials—some that they can pick up with tweezers, and some that they can't. Put all the choices on the table and let the children begin working.

Middle

To gain information on children's manual coordination abilities, watch the ways they use the tools. Do they hold the chopsticks and tongs in one hand, or hold one chopstick in each hand and use two hands to open and close the tongs? If they talk while they work, are they talking about how the tools work? ("When I squeeze and lift up, I can plop it in the bowl.") Or are they talking about some of the things you talked about in your introductory statement? ("Here are some things that work with the tweezers, but not with the tongs"). Look at the problems they encounter and the ways that they solve them—for instance, a pair of chopsticks may keep slipping from their hands when they try to pick up objects. When things like this happen, note whether they make several more attempts at picking up the object, ask another person for help, or simply move on to another item.

End

When the group time nears an end, as a cleanup strategy ask children to find three things on the table they can plop into the bowls. Ask children to talk about which tools they most enjoyed using for picking up objects.

Follow-up

1. Place the items used in a shoe box and add the box to the toy area as a work time option.
2. Bring a pair of chopsticks, a set of tongs, and a pair of tweezers to the next planning or recall time and have each child use one of these to bring an item back to the table that he or she plans to use (or did use) at work time.

Paper-Clip Designs

Originating ideas

For the past several days in the toy area, children have been working with put-together-take-apart toys, making them into long chains, then swinging them around and calling them "fishing poles."

Possible key experiences

Creative representation: *Pretending and role playing*

Movement: *Moving in nonlocomotor ways (anchored movement—bending, twisting, rocking, swinging one's arms)*

Classification: *Sorting and matching*

Seriation: *Comparing attributes (longer/shorter, bigger/smaller)*

Space: *Fitting things together and taking them apart*

Materials

✔ A collection of **paper clips** in a variety of sizes and shapes for each child, including traditional metal clips and new-style clips made of colored plastic

✔ Small **bowls or other containers,** one per child, to hold the clips

✔ For back-up: **pipe cleaners** or **twist ties**

Beginning

Pass out a container of paper clips to each child in the group and say "Today there are some clips inside your bowl for you to work with. See what different kinds of ways you can use them."

Middle

Look for all the different ways children combine the materials. Expect that some children will be fascinated by the different sizes and shapes of the clips and will sort them into piles based on these characteristics. Others will figure out how to clip them together and make them into chains and will spend the whole group time making and unmaking chains. Some children may attach labels to their work: "Look, I made a long chain"; "This is called my rope."

Imitate the various things you see children doing, using the materials in the ways children are using them. Occasionally make comments on children's work ("You put a lot on that one big clip"), or start conversations based on children's ideas (to a child who is pretending to catch fish, "You know, I went fishing this past weekend").

End

Place a large container in the center of the table as small-group time comes to an end. Say "Before you put your clips in the container, is there something you want to tell or show about the work you've done?" Listen to the children's explanations, repeating their comments and asking questions only if they are relevant to the children's work.

Follow-up

1. Bring the paper clips to the planning table the day after the small group, along with area symbol cards. Ask children to attach a paper clip to the area symbol card representing their work time choice.

2. Add the container of paper clips to the toy area near the other put-together-take-apart toys.

Teddy Bear Paths

Originating ideas

Recently, during the reading of the storybook *Where's My Teddy?* by Jez Alborough, children have been interested in the paths through the woods that are illustrated in the book. They like to run their fingers along the paths, saying "Forward, forward, back again, back again."

Possible key experiences

Seriation: *Comparing attributes (longer/shorter, bigger/smaller)*

Space: *Changing the shape and arrangement of objects (wrapping, twisting, stretching, stacking, enclosing)*

Materials

✔ **Paper strips** cut into different lengths and widths

✔ **Large sheets of construction paper,** one per child

✔ **Glue**

✔ **Teddy bear counters**

✔ A **sample path** to use for your introductory story (Make the path yourself using paper strips, glue, and a large sheet of paper.)

✔ For back-up: **crayons**

Beginning

Using the pre-made sample path and a teddy bear counter as props, tell a short story about a bear following a path into the woods and getting lost. Include descriptive language about shapes, positions, and sizes, such as, "First he walked over a narrow path, and then he came to a long road. By then he was very tired, so he was glad to see a very short road—but he was still lost." Ask children: "Can you tell the bear how to get back home?" Then act out their directions, using the bear as a prop ("Oh, you think he should walk backwards over the short road"). Pass out the materials and say "Here are some paper strips, paper, glue, and a bear for each of you to use to make up your own story."

Middle

Watch as children use the materials in their own ways, referring children to one another to encourage them to notice alternative ways to use materials. ("Your bear is walking down the long strip of paper. I notice that Leroy glued his short road on the paper so the bear doesn't have quite so far to go.") Go around the table and, using your own bear as a prop, ask children to give you directions on how to get somewhere. Look for children who use spatial language to give you directions ("Jump over the thick orange path"). Follow their directions with your own bear, making mistakes to see if they correct you (Plop the bear down on the orange path, and jump it over the blue path). For those children who are more interested in simply gluing paper strips onto their large sheet of paper, describe what you see them doing ("I see you glued a long, narrow strip next to the short, wide one"). If a child doesn't respond to your comment, move on to another child.

End

Ask some of the children to share their work with the whole group. Encourage children to put away the paper strips, glue, bears, and their creations in the appropriate places. Make a transition game by pretending the bear is once again lost and needs help finding his way to the next class activity.

Follow-up

1. Add paper strips to the art area. Announce their availability on the message board by attaching a paper strip or a child's creation from the previous day.

2. As a strategy for planning time, draw a map of the room with paths going to each of the classroom areas. As children tell their plans, have them use the map and a teddy bear counter as props, moving the bear along one of the paths to the interest area they plan to work in.

3. Take a piece of chalk outside and draw on the sidewalk some lines of varying widths. Play a game with the children that gives them the experience of jumping over the lines (or walking on the lines), encouraging them to notice and describe the different widths of the lines.

Spool-Print Rolling

Originating ideas

At work time and at outside time children have been rolling tennis balls, ping pong balls, and large rubber balls back and forth to one another. To provide a similar cooperative rolling experience in the art area, the teachers decided to introduce spools as art materials.

Possible key experiences

Creative representation: *Drawing and painting*
Initiative and social relations: *Building relationships with children and adults*
Movement: *Describing movement*

Materials

✔ **String-wrapped spools** in different sizes (You can use hard plastic, foam, or wooden spools. Prepare the spools for this activity by dipping pieces of string in white glue and wrapping the string around the spool in a random pattern, leaving most of the spool exposed. Allow the spools to dry before using.)

✔ One large sheet of **paper**

✔ Small **plastic containers** of various colors of **paint**

✔ Paint **smocks**

✔ For back-up: **smaller pieces of paper** for individual children to use

Beginning

Gather around the large sheet of paper with children. Dip the side of a spool into a paint container and tell children, "I'm going to see if I can roll this all the way across the paper. Let's see who it lands close to." When the spool comes to a stop, ask the child who is closest to roll it somewhere else. See if the children comment on how the spool makes a streak of paint lines as it moves across the paper. Pass out the rest of the spools so all children can begin painting.

Middle

Participate in the spool rolling and painting with the children, closely observing their inter-actions and their individual ways of painting and printing. Watch to see if they choose certain children to roll their spools too. Comment on what you see ("Adam, you and Avi are just rolling spools back and forth to each other, no one else"). Give them time to react to your words and add details ("Yeah, you're best friends today!"). If they print with the tops and the bottoms of the spools instead of rolling the spools, encourage them to describe their work by asking them questions ("I notice this mark is different than the others. How would I make one like that?"). Note which children decide to make their own design on a separate part of the paper rather than roll spools to others, and comment on this choice ("Today you're working alone").

End

As small-group time ends, ask children to describe how they will move their spools one last time. Have buckets of soapy water available so children can rinse the paint from the spools before putting them away. Working cooperatively with the children, lift the large sheet of paper from the floor and find a place for it to dry.

Follow-up

1. Add spools to the art area for children's use at work time.

2. Hang the cooperatively made spool picture at eye level with a real spool attached to it to remind children of the experience.

3. Do additional small-group activities that involve rolling objects and/or wrapping string around them. "Marble Painting," p. 88, and "Making Wood-Scrap Printing Tools," p. 142, are examples of such activities.

4. At planning time, choose the first child to plan by rolling a spool to the child. Then that child rolls the spool to the next child to plan, and so on.

Marble Painting

Originating ideas

The teachers have recently observed that children working in the toy area have lost interest in using the marbles with the plastic connector tubes. Following the previous small-group activity (see p. 86), string-wrapped spools were added to the art area, and children have been enjoying using these at work time. To provide a new kind of rolling experience, the teachers have planned an activity combining marbles with paint.

Possible key experiences

Creative representation: *Drawing and painting*

Movement: *Moving in nonlocomotor ways (anchored movement—bending, twisting, rocking, swinging one's arms)*

Music: *Exploring and identifying sounds*

Materials

✔ A **cardboard box,** lidded, containing a sheet of construction paper with a spoonful of wet tempera paint on it and two or three marbles

✔ Additional **boxes,** one per child

✔ **Construction paper** in a variety of sizes and colors

✔ **Marbles**

✔ **Paint** in **squeeze bottles**

✔ For back-up: **cotton swabs** or small **paintbrushes**

Beginning

Bring the prepared box to the small-group table, and without opening it, ask children to take turns shaking it gently from side to side. Ask if they can guess what is inside from the sounds made by shaking the box. After children make their guesses, open the box and show children the marbles and the "painting" inside.

Middle

In the center of the table, set out the rest of the boxes, the paint, the paper, and the marbles. Wonder out loud whether the box would make the same sound if it had two, or even three, marbles inside. Observe children's movements as they place the paper inside the box, add the paint, cover the box, then shake it. Are their movements coordinated? Do they have difficulty keeping the marbles inside their boxes, and if they do, how do they solve the problem? When they shake the box, do their whole bodies move, or do they plant their feet firmly on the ground and use only upper-body movements? Listen for the ways they describe the movements or the design the marbles have "painted" on their papers.

End

Before children put away their paintings and materials, ask them if they would like to share anything about their pictures with others at the table. Notice if they talk about the process of painting, the numbers of marbles they used, the colors, the way the marbles made lines across the paper, or the similarities their pictures have to others at the table.

Mixing marbles and paint inside tubs provides a new experience with marbles. In this version of the activity, teachers have provided open tubs instead of lidded boxes so children can see the effects they are creating.

Follow-up

1. At large-group time put on a fast musical selection (such as "Irish Washerwoman" from High/Scope's *Rhythmically Moving 3* recording), then a slow selection (such as "Rakes of Mallow" from *Rhythmically Moving 2)*. Ask children to pretend they are a marble inside a box being shaken to the beat of the music.

2. Add marbles and several boxes to the art area for future work time marble paintings.

Colored Water, Eyedroppers, and Suction-Cup Soap Dishes

Originating ideas

Children have been enjoying using eyedroppers, thin paint, and coffee filters at work time to create interesting designs and mixtures of color. Teachers planned this experience to introduce children to a new way of using eyedroppers to mix colors.

Possible key experiences

Language and literacy: *Describing objects, events, and relations*

Classification: *Exploring and describing similarities, differences, and the attributes of things*

Space: *Filling and emptying*

Materials

✔ Plastic **eyedroppers**

✔ **Rubber soap dishes,** the kind with many tiny suction cups (Look for these in the housewares department of discount stores.)

✔ Clear **plastic cups** filled with **blue, red, and yellow colored water**

✔ **Paper towels**

✔ For back-up: **Styrofoam trays**

Beginning

Set out the colored water, eyedroppers, and soap dishes. Say "Here are some soap dishes for you to use with the eyedroppers."

Middle

Observe the different ways children experiment with the materials. Some children will carefully fill each tiny suction cup on the soap dish with colored water, sometimes even making patterns with the colors. Other children will be fascinated by the way the cups stick to the table and by the noise made when they pull the soap dish back up. Listen to the ways children describe their actions or the materials they are using, and repeat their words or expand on

Experimenting with drops of colored water inside the tiny suction cups is an absorbing experience.

their thinking by posing new ideas. For instance, when this activity was tried in one preschool, one child called his work "making a chemical reaction" and another child said the eyedroppers were "pumper-squeezers." To encourage the first child to explain his description, the adult wondered aloud, "A chemical reaction —how so?" To support the second child, she tried imitating his pumping action on the eyedropper without putting it in the water, hoping this would encourage him to further explain his idea. Some children may only be interested in pouring and mixing the water in the clear plastic cups and watching the color changes.

End

Bring the paper towels to the table as the small-group comes to an end. Tell children you've brought materials they can use to soak up the extra water, and that a design may show up on their paper towels as they mop up the water. Working with the children, use the paper towels to clean up excess water, observing whether any children describe the changes they notice in the towels or anything else about the way the water is absorbed by the paper. Then have everyone collect the eyedroppers, soap dishes, and plastic cups and arrange them in separate piles.

Follow-up

1. Add soap dishes to the art area; store them near the eyedroppers.
2. Do the small-group time "Paper-Water Experimentation" (see p. 58).

Card Stacking

Originating ideas

There has been a lot of interest at work time in the interlocking Crystal Climber blocks, and these materials have been a factor in several child conflicts. The teachers have observed some children in tears because they haven't had enough Climber pieces to make large structures. Other children have gotten into fights over using the Climbers.

Possible key experiences

Initiative and social relations: *Building relationships with children and adults*

Classification: *Exploring and describing similarities, differences, and the attributes of things*

Space: *Fitting things together and taking them apart*

Materials

✔ **Small containers** for each child filled with old **playing cards** and **Crystal Climbers** (Cut 1-in. slits around the edges of the cards.)

✔ For back-up: **teddy bear counters**

Beginning

Pass out one Crystal Climber to each child, and have children cooperatively create a Crystal Climber structure by taking turns adding pieces. When everyone has had a turn, tell the children you have brought more Climbers for everyone to use, along with another material that has slits in it for putting together and taking apart.

Middle

Give everyone their containers of cards and Climbers. Look for the different ways they use them, and comment on their efforts. For example, some children may continue to build on the structure created at the beginning of the activity. In this case, the teacher might say something like "It's getting taller and taller." Other children may work alone, creating their own structures with their Climbers. Acknowledge these efforts by describing them ("Your building is wide with a whole lot of colors"). Other children may simply play with the cards, sorting the different suits into piles. Encourage these children to elaborate on their reasons for making groupings ("I notice you made piles. If I give you this card of mine, where will you put it?"). Still others may combine the cards with the Climbers, experimenting with fitting the Climbers into the slits on the cards.

End

Warn children that small-group time is about to end. Encourage those children who are interested to describe their efforts. Work with children to sort the cards and the Climbers into two separate containers in the middle of the table.

Follow-up

1. Store the slitted playing cards near the Crystal Climbers in the toy area for use at future work times.
2. Announce the addition of these materials on the message board, attaching an actual card and Climber to the board, along with your symbol for the toy area.

Cotton Ball Table

Originating ideas

Adults noticed that children rarely made plans to go to the sand and water table, which had been filled with sand since school started. To reawaken children's interest in the table, they filled it with cotton balls and planned this small-group time to introduce the new material.

Possible key experiences

Initiative and social relations: *Building relationships with children and adults*

Classification: *Exploring and describing similarities, differences, and the attributes of things*

Number: *Comparing the number of things in two sets to determine "more," "fewer," "same number"*

Space: *Filling and emptying*

Materials

✔ The **sand and water table,** filled with **cotton balls** and then covered

✔ A variety of **containers (bowls, pitchers, measuring cups)**

✔ A variety of **tools for scooping (ladles, large slotted spoons, tablespoons, measuring spoons, plastic cups)**

✔ For back-up: soft, cuddly **stuffed animals**

Beginning

Gather around the closed sand and water table with children. Tell them there is something new in the table and you will put some of it in their hands to feel and smell. Ask children to close their eyes, then put a cotton ball in each child's hands and see what words or actions they use as they explore the texture.

Middle

Lift the lid from the table and set out the additional materials. Notice which children continue to explore the texture of the balls by pulling them apart, twisting them through their fingers, squishing them down to see if they will flatten, and rubbing them against their skin to experience the softness. Observe which children begin to fill and empty the containers with the cotton balls. Play alongside children, imitating their actions as they explore the cotton and experiment with filling and emptying the containers. Occasionally describe your own actions or the actions of the children ("I wonder if I can fit as many cotton balls in this bowl as I did in the cup" or "First you filled a cup, then you filled a pitcher, and I noticed you used different tools to fill it"). Observe if children use the materials for pretend play; for example, they may offer one another cups of coffee, ice cream cones, and so forth. To support these children, take on a role in their play (sip the coffee, lick the ice cream).

End

Warn children when it is almost time to put the lid back on the table. Ask them to describe their favorite thing about having cotton balls in the table instead of sand. Enlist their help in carrying the lid over to the table and placing it back on.

Follow-up

1. Leave the cotton balls in the table as a work time option. Over time, observe how the cotton ball play evolves. Record the variety of ideas children have for incorporating cotton balls into their work time plans.

2. At planning time in the next few days, set out area symbol cards with clear plastic cups on each one. Give each child a cotton ball to put into the cup that represents an area he or she is interested in going to at work time.

Paper-Plate Decorating

Originating ideas

At large-group time in recent days, the teachers have planned several experiences in which children use paper plates to pat the steady beat of musical selections. The last time the plates were used in a movement activity, one child who was putting the plates away said, "I wish these could be pretty colors with streamers."

Possible key experiences

Creative representation: *Drawing and painting*

Seriation: *Comparing attributes (longer/shorter, bigger/smaller)*

At this large-group time, one of the children commented that she wished the paper plates they were using were more colorful. Responding to her interest, teachers planned the plate-decorating activity.

Materials

✔ **Large, white paper plates** for each child

✔ **Markers**

✔ **Colored streamers** of different colors and lengths

✔ **Materials for attaching the streamers** to the plates **(staplers, tape)**

✔ For back-up: **tape recorder** and **taped musical selections** (for example, "Alley Cat," from High/Scope's *Rhythmically Moving 3* recording)

Beginning

Start the group by saying that yesterday a child (name child) gave you an idea for a small-group time, something you could do with the paper plates that would make them colorful and fun to move with at large-group time. Put the materials on the table and say "See how colorful you can make the paper plates we'll use at large-group time."

Middle

To gain information on the amount of detail children use in representing with drawing materials, watch to see the ways children decorate the plates. Some will choose just one marker and cover the whole plate, front and back, with the same color. Others will use markers in a variety of colors, making various lines and shapes as they draw. Some children may make representations on the plates (a little girl dancing with plates). If children attach streamers to the plates, listen for their conversations or comments about the lengths, thicknesses, or colors of their selections. Imitate their actions, asking them for specific items to complete your own work ("Excuse me, Elyse—could you please pass me the thickest blue streamer so I can staple it to my plate"). When they tape or staple streamers to the plates, notice whether they do so independently or if they ask for adult or child support. Offer them help when needed ("You worked hard to staple the streamer, and it just doesn't seem to be working").

End

Give children a warning that small-group time is about to end. Help them put away unused materials in appropriate storage places, and set aside the decorated plates in a special location for later use at large-group time.

Follow-up

1. Use the paper plates at large-group time to tap to the beat of the "Alley Cat" selection from High/Scope's *Rhythmically Moving 3* recording.

2. Add streamers to the art area, along with paper plates in various colors.

Cooking Porridge

The other day at greeting time, the children asked about porridge as one of the teachers was reading them *Goldilocks and the Three Bears.*

Creative representation: *Recognizing objects by sight, touch, taste, and smell*

Seriation: *Comparing attributes (longer/shorter, bigger/smaller)*

Time: *Experiencing and comparing time intervals*

✔ Three large **bowls**

✔ Tools for stirring—**wooden spoons, tablespoons,** and **spatulas**

✔ **Instant hot cereal**

✔ **Milk** or **water**

✔ A **microwave oven** or **electric skillet**

✔ A **timer**

✔ **Ice cubes**

✔ Individual **serving bowls** and **spoons** for each child

✔ A **ladle**

✔ For back-up: **the storybook *Goldilocks and the Three Bears*** and several children's cookbooks in which the recipes are given pictorially

Beginning

Have children recall the part of the story in which Goldilocks eats the porridge. Note which children remember that the bowls were either too hot, too cold, or just right. Tell children you've brought all the materials they'll need to make their own porridge.

Middle

Make three separate bowls of hot cereal, letting children pour the dry cereal, mix in the milk, stir until the lumps disappear, and place the cereal in the microwave or skillet to cook. Heat the three bowls for different times (for example, for 40, 60, and 75 seconds), so that one bowl comes out "too hot"; one, "too cold"; and the last, "just right." As children wait, give them small amounts of the raw materials to explore. Make sure they have the option of tasting the "too cold" and "just right" cereals, and point out the steam rising from the "too hot" cereal. (As a safety precaution, add ice cubes to this bowl immediately after the children comment on the steam.)

End

Have children ladle their own servings of cereal, and eat alongside them as they enjoy the porridge they've made. When they've finished eating, ask them to rinse out their bowls and spoons and put away the materials used to make the porridge.

Follow-up

1. Add the empty cereal box and the other materials used for cooking to the sand and water table or the house area.
2. Recite the rhyme "Pease Porridge Hot, Pease Porridge Cold" at large-group time.
3. At large-group time, have children pretend they are Goldilocks eating the various bowls of porridge. Ask children what their reaction would be to the "too hot," "too cold," or "just right" bowls.

Working "Undercover" With Markers and Stamps

Originating ideas

The teachers have noticed that children have been using the stamps and stamp pads, as well as the markers, less often during work time. They planned this activity to see if providing these familiar materials with a new kind of work surface would rekindle the children's interest.

Possible key experiences

Creative representation: *Drawing and painting*

Language and literacy: *Writing in various ways—drawing, scribbling, letterlike forms, invented spelling, conventional forms*

Space: *Experiencing and describing positions, directions, and distances in the play space, building and neighborhood*

Materials

✔ **A large table** (such as the art area table)

✔ **Two large white sheets of paper or cloth,** one attached to the underside of the table and another attached to a nearby wall, so that children can draw on it if they are kneeling on the floor

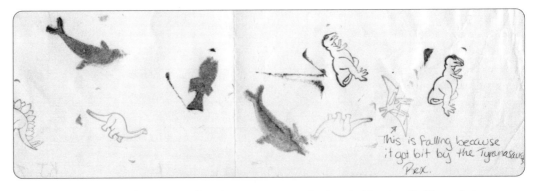

The teacher recorded this child's story about his stamp picture after he asked her to write his words down.

✔ **Markers**

✔ **Ink pads** and **stamps**

✔ For back-up: Individual **sheets of paper** for children who prefer to work alone

Beginning

Tell children that they will be using markers, stamps, ink pads, and paper for small-group time, and that they will be working in an unusual place. Give them some clues about the location of the paper or cloth. They will quickly notice the paper on the wall, so after the children guess that location, use it as part of the next clue ("That paper *is* easy to see, but the second piece is hiding *under* something").

Middle

After children have found the two work surfaces, put markers, ink pads, and stamps near the paper on the wall and under the table. Observe which children try drawing in both places, which have a preference for being under the table in tighter quarters with others, and which prefer drawing at the wall where the space is more open. Since this is a new drawing location, don't be surprised if children who normally draw recognizable representations instead draw with scribble marks. Notice how the parameters of the space define the children's drawings. Do they work with large sweeping arm movements at the wall and make smaller, more controlled writings under the table? Do they combine the stamp pads

Planning small-group time in an unusual location will sometimes encourage children to revisit familiar materials they have lost interest in.

with the markers to tell picture stories? Be available to listen to their stories, commenting or asking questions based on the text. Offer to take dictation from children who wish to add words to their pictures or picture stories ("I can write those words down if you would like").

End

After the small-group time ends and the materials are put away, gather back at your usual small-group meeting place and encourage children to talk about or show the positions their bodies took when drawing at the wall or under the table.

Follow-up

Leave the papers up. At planning time, ask children to find an item in the classroom that they plan to use at work time. Then have them trace around that item, using the paper on the wall or under the table.

Aluminum Foil Sculpture

Originating ideas

At work time, children have been wadding up newspaper from the art area shelf and taping it together to make three-dimensional designs.

Possible key experiences

Classification: *Using and describing something in several ways*

Space: *Changing the shape and arrangement of objects (wrapping, twisting, stretching, stacking, enclosing)*

Materials

✔ Pre-cut **aluminum foil pieces** in various lengths, shapes, and sizes

✔ **Crayons** and **markers**

✔ Materials for fastening—**paper clips, masking tape, string**

✔ For back-up: **Play-Doh**

Beginning

Bring to the table a wadded-up-newspaper creation designed by you or a child in your group. Tell the children that you have noticed them using newspaper to make designs that can stand up and that you have brought some more materials they can use today to crumple, stick together, and decorate.

Middle

Watch to see which children use the materials in two-dimensional ways, touching the shiny, smooth texture or drawing directly on the flat surface. Look for others who begin crumpling and building three-dimensional designs. Talk to them about the physical changes they are producing as they work on the foil ("It was flat and now it stands up"). Notice how children solve the problem of getting foil to stand up. Do they do it by manipulating the foil? Do they fasten pieces of foil together, using clips, wrapping the pieces with string, or taping them? As they work, watch for children who include others in their project, either through conversation about their ideas, or by requesting physical support to solve problems ("Please hold this here while I tie").

End

Ask children to describe some aspect of their work to the others. Point out the different ways children used the materials as a way of opening the discussion ("Micah, I noticed your paper is still flat, and Trey, your paper is standing by itself. Could you tell the others what your ideas were?"). When children have finished describing their efforts, store their materials so they can return to them at work time tomorrow or take them home.

Follow-up

1. Make a tray of aluminum foil pieces to put in the art area.

2. Add the empty aluminum foil box to the house area for role play.

3. Take the children to a park, shopping mall, or a museum where they can see what a sculpture looks like.

Glitter-Glue Designs

Originating ideas

One day, one of the children brought three glitter-glue pens to school with her and put them away in her cubby (adults have set the expectation that children should store materials brought from home in their personal cubbies). At work time, teachers noticed that a cluster of children had gathered around the child's cubby and were using the glitter pens to make designs on the floor.

Possible key experiences

Creative representation: *Drawing and painting*

Seriation: *Comparing attributes (longer/shorter, bigger/smaller)*

Materials

✔ **Glitter glue,** one bottle per child

✔ Colored **construction paper, white paper,** or **cardboard pieces** for children to decorate

✔ For back-up: **markers** and **crayons**

Beginning

Give each child a bottle of glitter glue and ask them to guess the secret ingredient. After they have guessed, set out the paper choices in the middle of the table so children can get started.

Middle

Notice which children explore the glitter glue rather than draw with it. As they explore, some children may focus on the sparkle the glue creates when it gets on their hands or other parts of the body, or they may simply dab glue or make a puddle of it on their piece of paper. Support children who are exploring glue by offering your own hand to receive a dab of glue or by crouching down at the child's level and joining him or her in looking closely at the puddle of glue on the paper. Note whether any children use the glue to make a simple representation (for example, a round blob that represents the sun) or one with details (a face with eyes, nose, and a mouth).

End

Ask children to put the glitter glue in the container you will use to store it in the art area. Show them the label you made for the shelf, and ask for their opinions about where it should go. Listen to see if they suggest storage ideas based on the attributes of the glue ("I think you should put it by the paper, because it goes on paper" or "Put it by the scissors because it's shiny").

Follow-up

1. Discuss the availability of the new material at greeting circle the day after your small-group time. Hang up a child's glue creation on your message board, so children who used the glue can explain to the others what the new material is.

2. Bring area planning sheets and glitter-glue bottles to planning or recall time, and ask children to "put a dot of sparkle" on the symbol for the area they plan to work in (or did work in).

Smell Collage

Originating ideas

Several of the children in the class have made up a game of guessing, from the smells in the air, what will be served for lunch. The game is becoming popular with other children in the classroom.

Possible key experiences

Classification: *Exploring and describing similarities, differences, and the attributes of things*

Materials

✔ A **large sheet of white paper** for everyone in your group to gather around

✔ An assortment of **spices—paprika, curry, ground cloves, whole cloves, ground cinnamon, stick cinnamon, nutmeg,** and **cumin**—and **Jell-O powder** (Place each material in individual bowls or shaker bottles.)

✔ **Glue**

✔ For back-up: Several **homemade "smell books"** created by gluing the above spices on individual pages of oak tag paper and adding word labels

Beginning

Gather around the paper with children and say "Today the things I brought for you to decorate this paper will also have some interesting smells, like the ones that come from the kitchen just before lunch time."

Middle

Participate with children as they get involved in smelling the different materials and gluing them to the paper. As you work, watch to see which children show interest in the smells and which use the materials simply to decorate the paper. Listen to the ways children describe the smells, and encourage them to tell you if the smells remind them of anything else from home or school. Make a guessing game out of searching for similarities or differences among the smells. If children begin eating the spices, encourage them to describe the tastes.

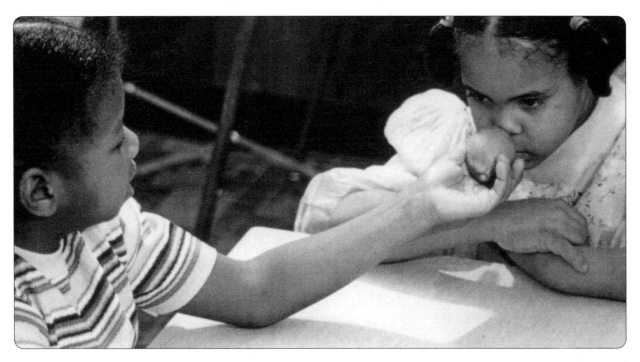

Smelling games like this one give adults clues about how everyday materials can stimulate children's thinking.

End

Shake off any excess spices from the collage
made by the group and put it in a safe place to dry. Have some children wash extra spices and
glue off the table while others use dustpans and brooms to sweep the floor. Decide on a place
where the collage will hang, being sure it is low enough so children will still be able to reach
it for sniffing.

Follow-up

1. Make a snack of cinnamon toast with the children.

2. Add the smell shakers to the art area.

3. Visit a local co-op or grocery store where spices are stored in bulk.

4. Take a spring walk to smell the flowers.

Sponge-Printing With Water

Originating ideas

Now that it is summer, teachers have been bringing the water table outdoors for outside time. Many children choose to go there, so there are often conflicts about space issues. The teachers have decided to introduce another way to use water (that does not involve the table) as an additional option for children who are interested in water play.

Materials

✔ **Large flat sponges** cut into a variety of shapes (footprints, hand prints, hearts, rectangles, squares, triangles, ovals, circles)

✔ **Buckets or other containers of water**

Possible key experiences

Creative representation: *Relating models, pictures, and photographs to real places and things*

Language and literacy: *Describing objects, events, and relations*

Movement: *Moving in nonlocomotor ways (anchored movement—bending, twisting, rocking, swinging one's arms)*

Classification: *Distinguishing and describing shapes*

Time: *Experiencing and comparing time intervals*

At the beginning of this small-group time, children take an active role by carrying containers of water outdoors to be used for sponge printing.

Beginning

Gather in your usual small-group meeting space. Ask children to find partners; then give each pair a container to fill with water and carry outside to a paved area, such as a sidewalk or asphalt play surface. When the group arrives outside, demonstrate what happens when you dip a sponge in water, squeeze it out, and then "print" with it on the pavement.

Middle

Join with children as they create shapes on the pavement. Watch their movements as they dip and squeeze the water from the sponges. Help them see the different attributes of the sponges, their shapes, their rounded or straight edges, the specific things they are intended to represent. Note whether any children observe that the prints evaporate faster in sunny spots, as compared to shady spots. If appropriate, comment on this difference yourself.

End

Have children find a sunny spot for their wet sponges so the sponges can dry before they return them to their storage containers. Play a hopping or jumping game as a transition to the next activity, using the water marks as the stepping stones to the next part of your routine.

Follow-up

1. Make sponges and containers of water available as a play choice for outside time.

2. Cut sponges used during other parts of your routine (for example, those used for washing tables before snack) into shapes like those used during the small-group activity. Observe whether any children notice and describe the similarities.

3. Draw several of each area symbol on a large sheet of paper. Then, at the next planning time, cover the symbols with dry sponges. When it is a child's turn to plan, he or she lifts up a sponge to uncover an area symbol. Then the child tells the others whether this area is involved in his or her work time plans.

Magnet Magic

Originating idea

At a recent planning time teachers hung magnets from the ends of homemade "fishing poles" and asked children to "fish" for the area symbol card of their work time choice (metal paper clips were attached to each card). The teachers noticed that some children stayed with the fishing game as their work time choice, laughing and calling it "magic" each time they managed to "catch" one of the "fish."

Possible key experiences

Language and literacy: *Describing objects, events, and relations*
Initiative and social relations: *Solving problems encountered in play*
Classification: *Sorting and matching*

Materials

✔ **Magnets,** one per child
✔ For each child, a **collection of small items,** some that the magnet will attract and others that it will not attract
✔ For back-up: **string**

Beginning

Pass out a basket of materials to each child. Say "There are some things inside your basket that will stick to the magnet, and some things that won't."

Middle

Give children ample time to explore the items the magnet attracts versus those it will not attract. Play with the magnet yourself, verbalizing the discoveries you see the children making ("I see the paper clip sticks to Adam's magnet"). After children have had enough time to explore, challenge their thinking by casually suggesting they try combining materials in new ways ("I wonder what would happen if we attached two or three paper clips together. Do you think the magnet could hold them all?" or "I wonder what would happen if we attached the piece of paper to the paper clip? Would the magnet pick up the paper then?"). Try the ideas yourself, but accept it if the children show no interest in trying your suggestions.

End

Working with children, put materials away in their original containers. Or, if children are older, encourage them to sort the materials according to whether or not they stick to the magnet. Store the materials in these groupings.

Follow-up

1. Add magnets to the toy area. Include other magnet games such as purchased magnet wands with magnet balls.

2. To make a transition to outdoor play, play a game in which children pretend their coats are magnets that are pulling them over to the coatracks.

Six-Pack Weavings

Originating ideas

Teachers have observed children weaving sticks into the chain link fence at outside time, so they decided to plan a small-group time to encourage another kind of weaving.

Possible key experiences

Movement: *Describing movement*

Space: *Changing the shape and arrangement of objects (wrapping, twisting, stretching, stacking, enclosing)*

Materials

✔ Plastic **six-pack rings,** several groups of six per child

✔ An assortment of **materials for weaving** through the six-pack rings **(strips of colored paper, straws, feathers, different lengths of yarn, pipe cleaners)**

✔ For back-up: **paper with pre-cut slits** to use for weaving

Beginning

Give each child a set of plastic rings from one six-pack. Pause for a few minutes to encourage children to examine the rings and talk about what they know about them. Some children may put the rings up to their eyes and call them eyeglasses, others may explain to you that the rings come from the tops of soda pop cans. Acknowledge the things children say. Then tell them the plastic rings remind you of the holes in the fence that you saw them putting sticks through the other day. Tell them that you've brought some other materials for them to use that will fit over and under the holes in the rings. Place the materials on the table.

Middle

Notice the different ways children combine the materials. Some may use the yarn to tie onto the plastic rings, making hanging ornaments or mobiles. Others may create weavings by stringing materials over and under the rings. Still others may twist and turn the pipe cleaners, but will not combine them with the plastic rings. Describe what you see children doing, using position and movement language, such as "twisting," "turning," "over," "under," and "through."

End

Encourage children to describe their work to others. Ask questions that encourage them to use spatial language ("How did you get this yarn to connect to the ring?"; "What were your hands doing when you put this pipe cleaner here?"). When it is time to clean up, encourage children to sort the materials back into their original containers.

Follow-up

1. Add extra six-pack rings to the art area, near the pipe cleaners, paper strips, and yarn.

2. Hang a sign near the parent information board requesting that parents save six-pack rings from home to bring to school. Fasten a single real ring to the sign to help children "read" the message.

3. Play over-and-under games at outside time by asking children to suggest places to run to ("What's a place we can run to that our bodies will fit under? Where our bodies can climb over?").

Shoe Box Decorations

Originating ideas

Observing that several of the children have been painting and gluing scrap paper to toilet paper rolls at work time, the teachers have decided to add other three-dimensional materials to the art area.

Possible key experiences

Creative representation: *Drawing and painting*

Initiative and social relations: *Making and expressing choices, plans, and decisions*

Classification: *Exploring and describing similarities, differences, and the attributes of things*

Materials

✔ A **shoe box** for each child with **glue** and **scrap paper** inside

✔ **Additional shoe boxes** in a variety of child and adult sizes

✔ **Paint** and **paintbrushes**

✔ **Markers** for the table

✔ **Smocks**

✔ **Newspaper** for covering the table top

✔ For back-up: **stickers**

Beginning

Tell children that you have noticed the way they have been decorating the toilet paper rolls at work time. Tell them you have brought a different kind of cardboard for them to use today at small-group time. Put one covered shoe box on the table, shake it, and ask the children for their guesses on what might be inside. After they have made their guesses and examined the materials inside, bring the newspaper to the table and enlist children's help in spreading it over the table. Then give each child his or her own shoe box to decorate.

Middle

Look at the ways children decide on the materials they will use to decorate their box, and comment on their choices. Do they use only one medium (paint only or glue and scrap paper only)? Do they combine materials and decorate both the inside and outside surfaces of the box? Do they choose materials based on materials another child at the table is using ("I'm painting like Brianna—She's my best friend")? Listen for the ways they describe and talk about their work to gain information on how they indicate their plans and intentions.

End

Ask children if there is something they want to tell you or their friends about their work before putting it away. Clear off the table top with the help of the children.

Follow-up

1. Bring a decorated shoe box to the next planning or recall time, and ask children to find a material in the classroom that fits inside the box that they plan to use (or did use) at work time.

2. Add extra shoe boxes and cardboard boxes in various sizes (jewelry boxes, cereal boxes, small shipping boxes) to the art area.

Enlisting children's help in cleaning up after small-group time helps to bring closure to the activity and encourages children to take responsibility for their actions.

Nuts and Bolts

Originating ideas

Recently, several children who were climbing on the park bench and outdoor climbing structure noticed and asked about the nuts and bolts that hold the metal and wood together.

Possible key experiences

Movement: *Moving in nonlocomotor ways (anchored movement—bending, twisting, rocking, swinging one's arms)*

Seriation: *Fitting one ordered set of objects to another through trial and error (small cup—small saucer/medium cup—medium saucer/big cup—big saucer)*

Materials

✔ Small **baskets,** one per child, containing a variety of different-sized **nuts and bolts** (at least six sets per child)

✔ For back-up: **narrow strips of thick cardboard with holes cut in them** to fit the nuts and bolts children have been given

Beginning

Pass out a basket to each child and say "Inside your basket are different-sized nuts and bolts. See if you can fit them together."

Middle

Since these are brand-new materials, children will probably be engrossed in exploring them and may not be interested in conversation. Therefore, spend most of your time noticing the ways children experiment with the materials and imitating their actions without comment. Some will lay the nuts and bolts out on the table, matching nuts to bolts one at a time. Some will begin fastening the nuts and bolts together, experimenting with screwing and unscrewing them. Others may notice that the nuts and bolts are similar to those on the outdoor climber and park bench and may talk about this connection. Still others may notice and describe how the different sizes do or don't work together ("This little one won't fit on this big one"). Respond to such comments by supporting further problem solving ("Do you see another nut that might fit better?").

End

Put a labeled container on the table for the nuts and bolts. As children finish working, watch to see which children put away their materials in the appropriate containers.

Follow-up

1. Ask children where they think the small-group materials should be stored in the classroom, and add the materials to the area (or areas) suggested.

2. Make a nuts-and-bolts board by drilling different-sized holes in a piece of wood, and fastening sets of nuts and bolts through the wood. With the board, store several wrenches that match the different sizes of the nuts and bolts, so children can gain experience with a new tool.

Making Colored Chalk Brighter

Originating ideas

Recently at outside time, the children have been playing with sidewalk chalk. Teachers have found a recipe for making chalk brighter and want to share the idea with the children.

Possible key experiences

Language and literacy: *Describing objects, events, and relations*

Classification: *Exploring and describing similarities, differences, and the attributes of things*

Materials

✔ Thick **sidewalk chalk** and thinner **blackboard chalk** in a variety of colors

✔ **Construction paper**

✔ One-half cup of **water** for each child

✔ Table **sugar**

✔ **Measuring tablespoons** and **stirrers**

✔ For back-up: **markers**

Beginning

Put the chalk in the center of the table and give a piece of construction paper to each child. After children have drawn some lines on their pieces of paper, bring the water, sugar, and other materials to the table. Say "I have materials to do an experiment that will make the chalk look different on your paper." Give each child who is interested a half cup of water. Put the bowls of table sugar near the children, and explain that you will be making a solution to dip the chalk into. Pass out the tablespoons, and tell children to put two tablespoons of sugar into their water, then to stir it.

Middle

Notice which children choose to experiment with the solution and which choose to continue drawing with the dry chalk. Look for children who notice and talk about the change in the chalk from dry to wet. Be available to expand on what children say with appropriate language, for example, use words like "dissolve," "brighter," and "wetter." Use your own set of materials to imitate the actions of the children.

End

Have a big bowl in the center of the table so children can discard their sugar water when they are finished. Find a place to store the wet and dry pictures. Sponge off the tables and expand on children's language, using words like "sticky" and "gritty" to describe the texture of the sugar water on the table.

Follow-up

1. Display the children's artwork at their eye level along with a simple picture recipe showing the water, the tablespoons of sugar, the sugar solution, and the chalk. For younger children, attach real items to the picture recipe.

2. Bring the sugar-water solution outside with large pieces of chalk so children can experiment with using the materials on the pavement.

Pounding and Hammering

Originating ideas

At outside time, children have been using large Tinkertoys as pretend hammers. They have been hammering everywhere in the outdoor play space—on the climbing structure, the pavement, the grass, the school building. As they work, they often make comments about "repairing" broken parts. Until this small-group time, children at this preschool had not used any real construction tools. Teachers felt that children's interest in hammering made this a good time to introduce real hammers.

Possible key experiences

Creative representation: *Pretending and role playing*

Movement: *Moving in nonlocomotor ways (anchored movement—bending, twisting, rocking, swinging one's arms)*

Materials

✔ Large **pieces of Styrofoam,** one per child

✔ **Lightweight hammers** and other **materials for pounding** (blocks, tubes), enough for each child

✔ **Plastic golf tees,** approximately 15 per child

✔ **Safety goggles** for each child

✔ For back up: **Play-Doh**

Beginning

Set a large piece of Styrofoam in the center of the table. Tell children you've brought materials for them to use for hammering golf tees into the Styrofoam, but before you give them their own materials, you will be pretending they have hammers in their hands. Together, act out the ways their hands will move in different situations (some in which hammers are used safely, some in which risks are taken), and engage them in conversation about what is safe and what is not. For example, ask them to imitate you as you make exaggerated motions with your own hands, raising them high over your head, and pounding down really hard on the Styrofoam. Then pose "what if" questions ("What could happen if you swing the hammer way back like this and someone is behind you?"). Then have them act out the smaller, more careful motions of safe hammering. After several of these play-acting scenes, give each child a large piece of Styrofoam. Ask children to find individual spots where they can work without being crowded by other children.

Middle

If needed, rearrange children so that each child is a safe distance from the others. Then give each child their golf tees and hammer. Move around to each child, commenting on the way the golf tee pushes into the Styrofoam when they hit it on the top. Since this is probably the children's first introduction to real hammers, watch them closely and comment positively on the safe ways they are using the tools ("You're keeping it down low, being careful not to hit your fingers—that's a safe way to use it"). If a child is not using a hammer safely, point this out, staying with the child until he or she is using the materials more carefully. Explain the consequences of unsafe use, including the result ("I'm worried that the way you are swinging the hammer will hurt someone"; "Swinging isn't safe, so if you want to use the hammer you'll have to stop swinging it"). If unsafe behavior persists, offer the child an alternative way of making holes in the Styrofoam (with just the fingers or with a block instead of a hammer). Support what you see children doing by imitating them. For example, pound randomly on the tees, hammer in a neat row of tees, or chip away at your Styrofoam as you work next to children who are using the materials in these ways. Echo and expand upon children's comments about their work. If children are having a conversation unrelated to pounding as they work, look for an opening and join in. Role play with children, following their lead about what your role should be ("Okay, I'll be your assistant and hold the golf tees while you hammer").

End

Give children a warning a few minutes before the group is about to end, using concrete language, such as "You have time to hammer in four more golf tees." Ask children to put their golf tees and Styrofoam pieces in containers or boxes you have prepared and labeled.

Follow-up

1. Set up a construction area in the classroom where children can use the Styrofoam pieces, golf tees, and safety goggles in their work time plans. Be sure the area includes a table top or other flat work surface.

2. Bring to the planning table a large piece of Styrofoam on which you have drawn the area symbols. Then give each child a chance to hammer a golf tee into the symbol that represents the area he or she will work in.

3. Add some big cut-off tree stumps to the outside area. Slowly, over several days, introduce real nails to children, a few at a time during outdoor play, making sure they wear safety goggles and reminding them of safety procedures.

III

Small-Group-Time Plans
Originating From

High/Scope
Preschool Key
Experiences

Working With Sticks, Straws, and Play-Doh Balls

Originating ideas

During a recent snack children compared short, thin pretzel sticks with long, thick pretzel sticks, commenting on their different thicknesses and lengths.

Key experience—Seriation: *Comparing attributes (longer/shorter, bigger/smaller)*

Possible additional key experiences

Creative representation: *Making models out of clay, blocks, and other materials*

Seriation: *Arranging several things one after another in a series or pattern and describing the relationships (big/bigger/biggest, red/blue/red/blue)*

Number: *Arranging two sets of objects in one-to-one correspondence*

Materials

✔ **Plastic bowls,** one per child, filled with **sticks, straws** cut into different lengths, and pre-rolled balls of **Play-Doh** in various sizes

✔ Separate, labeled **containers** for each of the materials, filled with extra pieces to use as needed during the small-group activity (These containers may also be used to collect leftover materials at the end of the activity.)

✔ For back-up: **pipe cleaners**

Beginning

Give each child a bowl of materials, and say something like this, "Here are some materials to work with—some are tall, some are short, some are round and big, and some are round and small. See what you can do with them."

Middle

Observe the ways children explore and arrange the materials. Some may use the straws and sticks to poke holes in the Play-Doh or may flatten or squeeze the Play-Doh balls with their hands or fingers. Listen for children's descriptions of their creations ("I made a lollipop"). Observe whether they use the materials to make a pattern (for example, alternating long and short straws) and whether they describe their patterns ("This is Daddy 'cause he's the biggest, Mommy 'cause she's the middle, and me 'cause I'm the smallest"). When children need additional materials, note the ways they ask for them ("I need another ball so this straw can have a top").

End

Set the containers of extra materials in the center of the table, and ask children to take apart their creations and return the materials to their places (the containers are labeled with picture or object labels). While they are cleaning up, note which children match materials with the appropriate container as they put them away.

Follow-up

1. Press a small ball of Play-Doh onto the end of a long straw and bring it to the planning table the next day, explaining to children that this is your friend Round Top. Ask children to share their work time plans with Round Top.

2. Put the containers of materials in the art or toy area for children to use in carrying out future work time plans.

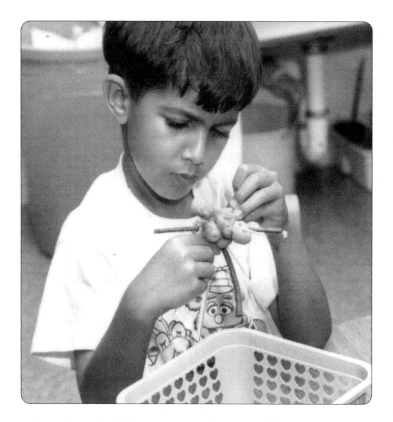

Adam makes a detailed representation of a person with stick, straws, and Play-Doh balls.

Shape Template Drawings

Originating ideas

At greeting circle several children have repeatedly requested the book Color Zoo by Lois Ehlert, which contains cutouts of various shapes that take the form of animals. While looking through the book, they have been tracing the outlines of the shapes with their fingers.

Key experience—Classification: *Distinguishing and describing shapes*

Possible additional key experiences

Creative representation: *Drawing and painting*

Initiative and social relations: *Solving problems encountered in play*

Movement: *Describing movement*

Materials

✔ The **book** *Color Zoo*

✔ Heavy cardboard or metal **shape templates** in various sizes

✔ **Construction paper**

✔ **Writing implements** (crayons, pencils, pens, markers)

✔ For back-up: **scissors**

Beginning

Bring the book *Color Zoo* to the group. Read the story to children, encouraging them to make comments or trace the shape cutouts with their fingers. As you read, take your cues from children, pausing longer on pages they have a lot of comments about and skipping others when there are no comments. Then tell children you have materials they can use to make their own shapes. Set the materials out on the table.

Middle

Choose a writing tool and a shape template and get a piece of paper. Begin tracing around the edges of the template or coloring inside it. As you work, notice the ways children approach the materials. Observe which writing implements they choose, and whether they trace around the edges of the template, fill in the whole shape, or make shapes on their own without using templates. Some children may not make any kind of drawing, but will explore the templates in some other way (for example, looking through the template at classmates, or tracing around it with their fingers). Observe whether children combine shapes to make representations on their paper (for example, using a triangular and rectangular body to represent the head and body of a person, or, as in the book, an animal). Notice the ways children solve the problem of keeping the template steady while they are tracing within it or around it.

End

Near the end of small-group time, ask children if any of the shapes they used remind them of something from the story you read at the beginning of group time. If they don't respond, wonder aloud whether there is anything they want to tell you or others about their work.

Follow-up

1. Use the shape templates at small-group time on another day, this time providing children with paints instead of markers. Then add the shape templates to the art area.

2. At large-group time, scatter the templates around the floor and have children dance around them to music. When the music stops, ask children to run to a shape and stand on it until the music starts again.

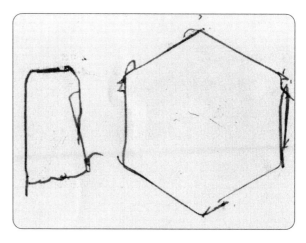

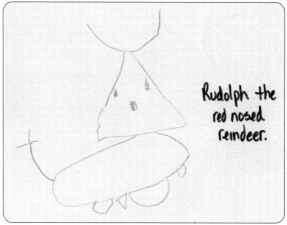

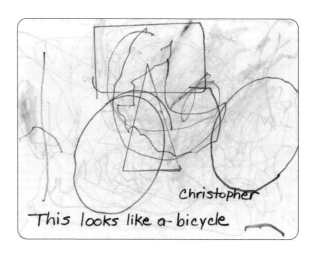

The artwork children make with templates reflects their different interests and ways of representing. Jake simply traced around the templates; Tamara traced around the triangle to make a head, then added details and other body parts to create "Rudolph, the red-nosed reindeer." Christopher traced around several shapes, and added some scribbles. Then he looked at it and said, "This looks like a bicycle."

Dried Beans and Egg Cartons

Originating ideas

During a recent work time a teacher observed two children placing single teddy bear counters and dinosaur counters on top of small, colored unit blocks. They arranged the unit blocks in various patterns on the floor, then put one bear or dinosaur on top of each block.

Key experience—Number: *Arranging two sets of objects in one-to-one correspondence*

Possible additional key experiences

Classification: *Sorting and matching*

Space: *Filling and emptying*

Materials

- ✔ A variety of **dried beans** (pinto, kidney, and lima)
- ✔ Plastic and cardboard **egg cartons,** cut into sections of two, three, or four cups each
- ✔ **Small boxes** or other containers, one per child
- ✔ **Three prepared containers** for collecting and sorting the beans at cleanup time, one for each variety of bean—include a pinto bean container labeled by gluing a single pinto bean to the outside of the container, a kidney bean container labeled with a drawing of a kidney bean, and one unlabeled container for the limas
- ✔ For back-up: **playing cards**

Beginning

Say something like "I put some beans and egg carton cups inside each container today for you to use at small-group time. See what you can do with them." Then give each child a container filled with beans and several of the egg carton sections you have prepared.

Middle

Observe the many ways children use the materials. Some children will probably begin sorting the different types of beans into piles or into separate egg cups. Others may use the egg carton sections for filling and emptying—you may observe these children filling the egg carton cups with a random assortment of beans, and experimenting with pouring them back and forth between the sections. As you move around the table, listen to and converse with children, imitating their actions when appropriate.

End

As small-group time draws to a close, put the three prepared bean containers in the middle of the table. Tell children these containers are for putting away their beans, and observe to see whether children sort according to the labels. Note whether anyone suggests that the lima beans go into the unlabeled container.

Follow-up

1. Place beans and egg cartons in the toy area.
2. At planning time, give each child a turn to place a bean on an area symbol card representing the area they plan to go to at work time. Then have the child describe his or her plans for the others.

Teddy Bear Stories

Teachers have noticed that during morning greeting time, children frequently request the book Where's My Teddy? *by Jez Alborough.*

Key experience—Language and literacy: *Having fun with language: listening to stories and poems, making up stories and rhymes*

Possible additional key experiences

Creative representation: *Making models out of clay, blocks, and other materials*

Materials

✔ The **storybook,** *Where's My Teddy?* by Jez Alborough

✔ Small- and large-sized **teddy bear counters**

✔ Little **plastic people** and **inch-cubes or Cuisenaire rods**

✔ For back-up: a variety of **small boxes**

Beginning

Find a comfortable spot to meet where children can see and actively participate in the reading of the story. Read the book to them, stopping often to give them time to comment *(do not make comments such as "Story time is my turn to talk and your turn to listen" or "We'll answer your questions after the story")*. Respond to and discuss any comments children make. For example, in one classroom, a child responded to this sentence from the story—"The sound of the sobbing comes nearer and nearer"—by asking, "What means *sobbing?*" As you read, invite children to share their own experiences with teddy bears, real bears, or with being in the woods. Encourage them to interpret the feelings of the characters by focusing on their facial expressions. When you have finished reading the story, gather with children at the small-group table and pass out the additional materials.

Middle

Tell children you've brought materials so they can tell their own story about a teddy bear and its friend. As they begin to work, move from child to child and watch to see if they incorporate words and text from the original story. Listen also for comments and ideas that represent a new teddy bear story. Look at the ways children use materials to make representations. Some will use them to make the path in the woods or the beds the characters rush to at the end of the story. Others will simply make stacking towers out of the blocks or Cuisenaire rods, while others will sort blocks and teddy bears into color-matched pairs.

End

Move around the table holding your own teddy bear, reminding children that in just a few more minutes, the teddy bears, people, and blocks will have to be put back in their containers. Ask children (speaking through your teddy bear), if they would like to share anything about what they just did.

Follow-up

1. Bring the area symbol cards and the teddy bear counters to planning on the day after the small-group time. Give each child a bear, and ask the children, one by one, to put their teddy bear on the area card that represents the place they will begin their work time plans.

2. Add *It's the Bear* (a sequel to *Where's My Teddy,* with the same characters, but older—also by Jez Alborough) and *Little Polar Bear* (a story about a different kind of bear, by Hans de Beer) to the book area.

Using Toothbrushes

Originating ideas

Because licensing regulations require children to brush their teeth daily, toothbrushing is a part of the routine in this program. During these toothbrushing times, teachers have observed that children are at different levels in understanding why and how to brush. Since toothbrushes must be replaced at regular intervals, small-group time provides an ideal opportunity for adults to introduce new brushes and to observe children's brushing more closely.

Key experience—Initiative and social relations: *Participating in group routines*

Possible additional key experiences

Initiative and social relations: *Taking care of one's own needs*

Time: *Anticipating, remembering, and describing sequences of events*

Materials

✔ **Toothbrushes** and **tubes of toothpaste,** one per child, each clearly labeled with his or her name and symbol

✔ **Cups,** one per child; several **small pitchers of clean water,** and a **large container** to fill with used water

✔ A **full-length mirror** (you may have one in the house area) turned sideways on the table

✔ A large set of **plastic teeth** (this is optional—you may be able to borrow these from a dental or public health office) and a **large toothbrush**

✔ Four or five **hand mirrors**

✔ For back up: **dental floss**

Beginning

Using the plastic teeth and large toothbrush as props, tell a story about a set of teeth that just finished eating but forgot how to brush. (If you don't have plastic teeth, use your own teeth as props.) As you tell the story, ask children for their ideas on how to help the teeth. Build the content of the story around children's ideas, being sure to include health-related issues (using your own brush, spitting water into your own cup) as they naturally fit with the story. After the story ends, gather around a table on which you have arranged the individual toothbrushes, toothpaste, and mirrors. As children begin using the toothbrushes and toothpaste, add the pitchers of clean water and the container for the waste water.

Middle

Watch the ways children approach the task of brushing their teeth. Observe which children independently squeeze their own toothpaste onto the brushes and which ask for help. Look for children who simply pour water from various containers, those who use the mirrors to examine their mouths or other parts of their bodies, and those who use the mirrors while brushing their teeth. Listen for children who give verbal instructions to others on the steps in brushing. If you notice children are brushing incorrectly, avoid making this an issue, except in a playful way. (For example, if you notice a child brushing only the front teeth, you might say something like "'Ouch' say the back teeth, 'brush me too!'")

End

Have children put away their toothbrushes so they are stored in compliance with state or local health regulations. Put away hand mirrors and enlist children's help in returning the long mirror to the house area.

Follow-up

1. Use old toothbrushes for painting in an upcoming small-group time (see "Toothbrush Mural," p. 32). Put a brightly colored piece of tape on each of the old toothbrushes to give children a visual reminder that these toothbrushes are to be used for painting rather than brushing teeth.

2. Add *The Berenstain Bears Visit the Dentist* by Stan and Jan Berenstain to the book area.

As a follow-up to the toothbrushing activity, teachers provide old toothbrushes for children to paint with at another day's small-group time.

Stapled Paper-Scrap Designs

Adriana and Saraya spent all of work time arranging and rearranging their artwork on the floor in the book area. (Children's artwork is kept in a box located next to the entrance, near the book area, so parents can easily get children's projects to take home.) As they worked on their art display, the children used positional phrases like "above the blue one," "under mine," and "next to yours."

Key experience—Space: *Experiencing and describing positions, directions, and distances in the play space, building, and neighborhood*

Possible additional key experiences

Initiative and social relations: *Making and expressing choices, plans, and decisions* and *building relationships with children and adults*

Movement: *Moving in nonlocomotor ways (anchored movement—bending twisting, rocking, swinging one's arms)*

Materials

✔ **Large sheets of plain white paper**

✔ For each child, a collection of colored **paper scraps** in many different sizes and shapes

✔ Enough **staplers** so each child can have one

✔ Several rolls of **clear tape**

✔ For back-up: **markers**

Beginning

Gather around the table. Say "Yesterday, I saw Adriana and Saraya laying out papers in different ways. Today I brought some small scraps of paper to see how you might use them." Give each child his or her own basket of paper scraps and piece of paper. Put the staplers and tape in the center of the table so children, if they choose to, can make their own designs of stapled or taped paper scraps.

Middle

Watch the ways children arrange and rearrange their scraps and listen for any descriptions they give of what they are doing. See if any children pair up to collaborate on a design. Watch children's movements as they staple or tape their work. Do they easily manipulate the tools, or do they ask for an adult's or another child's help in holding the paper steady or in pushing down the stapler?

One day at work time, Saraya and Adriana went to the box where chil-dren's artwork is stored and spent all of work time arranging and rearranging their artwork. This small-group time is designed to build on children's interest in making spatial arrangements.

End

Ask a few children to share their designs. Create a guessing game by giving clues that involve spatial language: "I see someone with a yellow shape sticking out of the side of his work. Is there anything you want to tell us about your little pieces of paper?" Continue the game, encouraging children to give the clues.

Follow-up

1. Add the story *Hop on Pop* by Dr. Seuss to the book area.
2. Put a box of paper scraps in the art area.

Masking Tape Magic

Children have recently been playing a made-up game. They pull long strips of masking tape from the roll, crumple it up, then throw the ball of tape at the wall. If it sticks, they say, "Look, it's magic."

Key experience—Classification: *Using and describing something in several ways*

Possible additional key experiences

Creative representation: *Drawing and painting*

Language and literacy: *Describing objects, events, and relations*

Seriation: *Comparing attributes (longer/shorter, bigger/smaller)*

Number: *Arranging two sets of objects in one-to-one correspondence*

Materials

✔ **Cardboard** or **mat board** paper
✔ Several rolls of **masking tape**
✔ **Markers** or **crayons**
✔ For back-up: **scissors**

Beginning

Show the children a piece of cardboard with a piece of masking tape attached to its center. Lay the cardboard in the center of the table so that all the children can see it well. Tell them you have a new way to make magic, and rub a crayon back and forth across the tape. Ask a child to pull off the tape, then listen for children's comments. Give each child his or her own piece of cardboard with several pieces of masking tape on it. Set the other materials out on the table.

Middle

Notice whether any children choose to continue with the crayoning-on-tape technique you have just demonstrated. Some may decide not to use the tape with the cardboard or the markers, and instead will repeat their made-up game of throwing balled-up tape against the

One child's artwork illustrates the interesting effects created when the masking tape strips are removed.

walls. Others may simply decide to draw and write with the markers on the cardboard pieces. Others may use the tape—drawing or coloring on top of it, but leaving it attached to the paper. Still others may add new pieces of tape and crayon over those. Listen for the ways the children describe their actions, and notice if any continue to use the word *magic* and create new definitions of the word.

End

Tell children when it is almost time to clean up and give them the opportunity to take their creations home or display them in the room.

Follow-up

1. Put a number of pieces of masking tape on a large sheet of cardboard or tag board. Draw several representations of each area symbol over the masking tape, so that part or all of each symbol is drawn on the tape. At recall time ask each child, in turn, to get an item he or she used at work time. When a child returns from an interest area, ask him or her to find a symbol for that area in the masking tape picture and to pull off one of the pieces of tape underneath the symbol. Then the child can talk about what he or she did with the item in the interest area.

2. To give children an experience with another kind of magic, add *Sylvester and the Magic Pebble* by William Steig to the book area.

Doctoring

| Originating ideas |

During work time children have been announcing they are "dead" or "hurt." Other children are pretending to "operate" on these patients, wearing heavy gloves and using scarves to represent the bandages and casts they are using to "fix" their patients.

Key experience—Creative representation: *Pretending and role playing*

| Possible additional key experiences |

Initiative and social relations: *Solving problems encountered in play*

| Materials |

✔ A collection of **baby dolls** and **stuffed animals,** enough so there is one for each child in the group, with a few extras so that each child has a choice

✔ Strips of **cotton cloth**

✔ Wooden **dowels or flat sticks**

✔ Various **medical materials and toys: Band-Aids, Ace bandages, toy syringes, plastic gloves,** and **toy stethoscopes**

✔ For back-up: **blankets**

Beginning

Bring a bag filled with stuffed animals and dolls to the small-group table. Pretend you hear crying sounds from inside, then pull a doll or stuffed animal from the bag and put it to your ear. Tell children the doll whispered to you that she/he is sick. Bring out the other medical materials and enlist children's ideas on the ways to help the sick doll or animal. Act out some of the children's ideas, then tell them you hear more children and animals crying in the bag. Give each child a chance to select a doll or animal to "fix."

Middle

Watch to see which materials the children select as they work on their patients. Do they repeatedly give shots with the syringe? Do they put on gloves before working on a "patient"?

Are they fascinated with the motions involved in opening up and applying Band-Aids? Do they enlist the help of others as they wrap cloth around splints? As they work, listen for their language. Note which children comfort their patient ("Don't worry, you'll be better soon"; "It only pinches for a moment") and which children describe their actions ("Okay, I'm gonna poke you with the shot now, so don't move"). "Fix" your own doll or animal, and offer assistance to children who ask for it. Encourage children who lose interest in their patients to "doctor" you or a willing friend.

End

Tell children they have 5 more minutes to work on their "patients." Go around and ask children's dolls and stuffed animals if they feel better. ("Oh, hi, kitty, I noticed Asia worked hard to help you today. Do you feel better?") If the children respond to your effort, listen carefully to their explanations.

Follow-up

1. Add medical equipment to the house or block area.
2. Put *Curious George Goes to the Hospital,* by H. A. Rey, in the book area.

Brittany bundles up her sick teddy bear and gets ready to feed him a bottle.

Flashlight Tag

The teachers have observed children running up and down a hill, climbing, and jumping at outside time, so they planned this small-group time to provide another kind of large-motor activity.

Key experience—Movement: *Moving in locomotor ways (nonanchored movement—running, jumping, hopping, skipping, marching, climbing)*

Possible additional key experiences

Creative representation: *Imitating actions and sounds*

Initiative and social relations: *Making and expressing choices, plans, and decisions*

Space: *Observing people, places, and things from different spatial viewpoints*

Time: *Starting and stopping an action on signal*

Materials

✔ A **flashlight**

✔ A **bag** containing **cards with the personal symbols** of all the children in the group

✔ For back-up: Large strips of **elastic**

Beginning

Find a large gathering place, for example, around a tree outdoors or in the corner of a large space indoors. Tell children they will be playing a game with flashlights that involves moving with their whole bodies. Hold out the bag containing children's personal symbols, and ask a child to choose a symbol from the bag. The child whose symbol is chosen is the first to hold the flashlight. Ask that child to shine the flashlight in a certain place, then describe a way to move to that place (for example, running, climbing, hopping, marching, crawling or jumping). After everyone arrives together at the same location, another person takes a turn. Continue to have children choose from the bag of symbols to pick the child whose turn comes next.

Middle

Continue to play the game, following the ideas suggested by the children, until everyone has had at least one turn to point the flashlight toward a particular location. Encourage children to describe the movements they have chosen, and notice which ones use movement words and phrases such as "hopping," "jumping," "twirling all around." Each time the group reaches a child's chosen destination, take a moment to point out the distance traveled and the things you've passed along the way. Then ask another child to choose a destination.

End

After all children have had a turn, tell them you will choose two more symbols, and that after those turns, small-group time will be over (this will help to eliminate some of the focus on waiting for another turn). After the children selected have had turns, ask children to move to the area for the next part of the daily routine in any way that they choose.

Follow-up

1. Bring a flashlight to planning or recall time for children to shine on or towards an area or object that figures in their work time plans or was part of their work time experiences.

2. Add flashlights to the materials available for children to use at work time.

Making Wood-Scrap Printing Tools

Originating ideas

This activity is a follow-up to the spool-printing activity presented on page 86. During that activity, children were curious about how the string got stuck on the spools—they had never seen spools like this before. To respond to children's interest, teachers planned this small-group time, in which children have the opportunity to create their own string-tied objects to use for printing.

Key experience—Creative representation: *Making models out of clay, blocks, and other materials*

Possible additional key experiences

Space: *Changing the shape and arrangement of objects (wrapping, twisting, stretching, stacking, enclosing)*

Materials

✔ **Pie tins** containing **pieces of string** approximately 12 inches long, soaking in a mixture of equal parts of **glue** and **water**

✔ Additional pieces of **dry string** for children who want to dip their own pieces

✔ **Wood scraps**

✔ A **string-wrapped spool** from the spool-printing activity in a **paper bag**

✔ For back-up: **paintbrushes**

Beginning

Put one of the spools from the spool-printing activity in a paper bag and bring it to the table. Ask each child in turn to put one hand inside the bag to feel the contents and then to guess what is inside. As children make their guesses, encourage them to elaborate on the reasons for their ideas. After you pull the spool out of the bag, tell children you have brought wood scraps and wet, sticky string so they can make their own printing tools.

Middle

Join children in wrapping the glue-soaked string around the wood scraps. Watch as they solve the problem of holding a wood scrap in one hand while wrapping string around it with the other. Listen for their comments about the string, observing whether they notice and describe the differences between the glue-soaked string and the dry pieces. Note any words they use to describe the differences. Look at the ways the children cover the wood. Some will wind one piece of string around a wood scrap and say they are done, and others will apply many layers of string to their piece of wood. See if any children compare their wood pieces with the spools used in the other activity.

End

Store wood scraps on drying racks instead of on newspaper to avoid problems created by sticky string. Wash the glue-water mixture off the pie tins and make a space for them near the glue on the art shelf, next to any leftover pieces of dry string.

Follow-up

1. To announce to children that the materials from this activity are available in the art area, attach one of their string-wrapped wood creations to the message board, along with an arrow pointing to the art area symbol.

2. At planning or recall time, give each child a planning or recall sheet with area names and symbols on it. In the middle of the table, set a tin of paint and one of the string-wrapped wood scraps that has been allowed to dry. Ask children to dip the wood-and-string creation into the paint, then use it to stamp on the area symbols representing the area(s) where they plan to go (or went) at work time. As they work, talk to them individually about their plans. Don't be surprised if some children find this printing activity so engrossing that they choose to continue to do it at work time.

Paper Bag Stuffing and Decorating

Originating ideas

Around Halloween, large pumpkin- and cat-shaped leaf-collection bags are available in many local stores. Teachers brought several of these bags to the classroom, and the children decided to fill them with leaves and crumbled newspaper. After filling the bags, children asked if they could take them home to decorate their own houses or yards.

Key experience—Creative representation: *Making models out of clay, blocks, and other materials*

Possible additional key experiences

Space: *Changing the shape and arrangement of objects (wrapping, twisting, stretching, stacking, enclosing)*

Materials

✔ **Paper bags** in varying sizes (grocery, lunch, and shopping bags)

✔ **Newspaper**

✔ **Tape, string,** and **scissors**

✔ **Paint** and **paintbrushes**

✔ For back-up: **glue** and **paper scraps**

Beginning

Bring one of the plastic bags from outdoors to the table. Using the bag as a puppet, talk to the children about their request to take home the bags like it. Explain through the voice of the puppet that the bags belong to the school ("I know you'd like to take me home, but my home is here at school"). In the puppet's voice, point out that there are materials on the table children can use to make their very own bags to take home.

Middle

Notice the ways children combine and use the materials. Some will draw or paint on the flat paper bags. Others will crumple newspaper and fill their bags with it. Still others will fill bags and use tape or string to fasten them closed. Observe the different ways children solve the problem of fitting the newspaper inside their bags and closing off the tops. Be careful not to suggest that children make cats or pumpkins, looking instead at the ways the materials inspire children's own ideas. Some children *will* use the bags to make stuffed pumpkins or cats; others may make other types of models or puppets; still others may use the bags to make balls, which they will then toss up and down. Listen for children's explanations of how they will use their creations.

End

When children are finished, line up children's paper-bag creations near the door, being sure to write children's names or symbols on each. With the children, return paint, tape, and scissors to their places and wash the table tops.

Follow-up

1. The day after this small-group time, remind children at planning time about their paper-bag creations, in case they want to add details or make more.

2. Add newspapers and paper bags to the art area and announce this on the message board at greeting time.

3. Before recall time, make a ball by stuffing crumpled newspaper inside a bag and fastening it shut. Bring the ball to the recall table. Ask the child who is describing work time experiences to throw the ball into the area he or she worked in (or toward a material they used). Have the child retrieve the bag along with an item used at work time and then come back and talk about what he or she did with the item.

Outdoor Slime Racing

In a previous small-group time, children make slime that was used in this active outdoor activity

Each time teachers have provided opportunities for children to run with objects at outside time, children participate enthusiastically. Teachers planned this activity to enable children to experience running with a very different kind of object. The children are familiar with the "slime" used in this activity, having made slime in the previous day's small-group time (p. 28).

Key experience—Movement: *Moving in locomotor ways (nonanchored movement—running, jumping, hopping, skipping, marching, climbing); Moving with objects*

Possible additional key experiences

Initiative and social relations: *Solving problems encountered in play*

Movement: *Moving with objects*

Space: *Filling and emptying*

Time: *Starting and stopping an action on signal*

Materials

- ✔ Three or four big **bowls of slime** (for recipe, see p. 28; for added fun, make slime in several colors.)
- ✔ Three or four other **large, empty containers** scattered around the play space (Place the empty containers at moderate distances from the slime-filled containers so children have to solve the problem of how to get the slime from the full containers to the empty ones.)
- ✔ **Smocks**
- ✔ For back-up: **large spoons or ladles**

Beginning

Have everyone put on smocks and gather outdoors around the full bowls of slime. Ask children to recall what happens with slime when you pick it up, referring to the previous day's experiences. After you have discussed this, point out the empty buckets scattered around the play yard and tell them that for this small-group time they will be moving the slime from the full containers to the empty ones, using only their hands.

Middle

As children move slime from full to empty containers, expect that some slime will fall on the ground. Don't be concerned about this—the slime is easy to clean up. Watch for children's reactions when they realize that slime will escape from their fingers before they reach the empty bowls. Do they walk or run faster? Do they work out cooperative strategies for moving the slime? Do they move the bowls closer? Do they reach down and try to pick up the slime? Do they stand and watch others move slime without participating themselves? Don't be surprised (or try to change their actions) if some children stay close to a single container, exploring the material without running. As children experiment with the slime, play alongside them, staying focused on the goal of moving slime from bowl to bowl. Model new ways of getting slime from place to place, for example, passing some from your own hands to another child's. Follow children's cues. For example, if children decide to make a game out of throwing slime to one another, participate with them.

End

To clean up, bring out hoses or buckets of water. Slime will dissolve into the grass and disappear from children's hands when water is added.

Follow-up

Gather additional materials children can use for playing games involving running, filling, and emptying (for example, cotton balls, plastic eggs, pebbles, acorns, chestnuts). In your outdoor play area, set out buckets of these materials, along with additional empty buckets and things to use as scoops.

Tire-Tread Paintings

At outside time children have been commenting on the different track marks the tires of the trucks and cars make in a dirt pile that is next to the school parking lot.

Key experience—Creative representation: *Recognizing objects by sight, sound, touch, taste, and smell*

Possible additional key experiences

Seriation: *Comparing attributes (longer/shorter, bigger/smaller)*

Materials

✔ A collection of **plastic cars and trucks** with different tire sizes and textures

✔ **Pie tins** filled with **paint**

✔ Standard-sized sheets of **paper**

✔ Several **buckets of warm, soapy water**

✔ **Sponges**

✔ For back-up: A **large piece of paper,** big enough for children to work on cooperatively

Beginning

Beforehand, use paint, paper, and two of the toy vehicles to create two distinctly different sets of tire tracks on a piece of paper. As small-group time begins, show children the two vehicles and the paint prints, and ask children to guess which car or truck made the marks. Give each child a set of materials consisting of three or four vehicles, a pie tin filled with paint, and a piece of paper.

Middle

As children begin to use the materials, watch to see if they make deliberate choices about cars and trucks with different tire widths. Listen for any sounds they make as they draw. (Do they make car or truck noises as they move the objects across their paper? Do they sing songs as they work?) Make your own picture, making marks like those of a specific child near you. Ask the child to pass you the object he or she has used to make the marks. Observe whether the vehicle the child hands you actually does match the marks; of course, don't correct a child who hands you a different vehicle.

End

When children have finished painting and putting away their work, set out buckets of water and sponges, encouraging children to dip their vehicles in the water and wipe off the paint. Join in this cleanup process by washing several vehicles of your own.

Follow-up

1. Bring a clump of Play-Doh to the planning table. Ask the children to get an object they will use at work time and then to make an imprint with it by pressing it into the Play-Doh. As each child discusses his or her object and plans for work time, ask the other children to guess what part of the child's object was used to make the imprint. Then have the planner talk about how he or she plans to use the object.

2. With children, make up a song at large-group time about the tire-track painting activity. Use the tune from "The Wheels on the Bus," and give children the opening phrase, "The tracks on the car. . ." Encourage children to fill in the rest of this verse and to suggest additional verses.

Wrapping Paper Decorating

Originating idea

During the past several work times, Kacey, Julia, and Emma have played a pretending game about Santa Claus. In this game they first wrap up toys in newspaper and tape. Then, they pile them next to Julia, who sits in a chair with cotton taped to her chin. Then Emma and Kacey call to the other children, telling them Santa is at the mall and ready to give them presents.

Key experience—Language and literacy: *Talking with others about personally meaningful experiences*

Possible additional key experiences

Creative representation: *Drawing and painting*

Language and literacy: *Describing objects, events, and relations*

Classification: *Exploring and describing similarities, differences, and the attributes of things*

Materials

✔ Plain pieces of white **paper**

✔ **Cookie-cutters** associated with the children's holiday role-playing (such as holiday trees, stars, and bells), plus several non-holiday-related cookie cutters (dogs, houses)

✔ **Paint tins** to dip the cookie-cutter shapes in

✔ A brightly colored piece of **holiday wrapping paper**

✔ For back-up: **glue** and **glitter**

Beginning

Bring a newspaper-wrapped item from the children's Santa play to the table. Ask Julia, Kacey, or Emma to talk about what they have been doing with newspaper and toys during work time. After they have had a chance to describe their present-wrapping activities, bring out a brightly colored piece of holiday wrapping paper and encourage children to contrast it to the newspaper wrappings. Listen carefully to their comparisons and descriptions. When children are finished sharing their ideas, show them the white paper, the paint, and the cookie-cutters. Tell them you've brought materials they can use to make wrapping paper or a picture that is bright and colorful.

Middle

Join in as children begin dipping the cookie-cutters into the paint and printing. Notice which children carefully make patterns and designs on their paper, separating shapes from one another, and using a variety of paint colors. Watch which children use the cookie-cutters like paintbrushes, dipping them in the paint, then spreading the color over the entire surface of the paper. Take part in children's conversations about the shapes of the cookie-cutters, their experiences with giving and receiving gifts, and other holiday experiences they may bring up.

End

Have a bucket of warm, soapy water available so you and the children can clean the cookie-cutters as the small-group time comes to an end. Ask children if they want to keep the decorated paper in the room, and if so, show the children where it can be stored in the art area.

Follow-up

1. When the wrapping paper is dry, have children put it away in the suggested spot.
2. Use some of the decorated paper and newspaper to wrap items you saw children using at work time, and bring them to the recall table. Have a child unwrap an item and guess who used it. Then have that child recall his or her experiences.

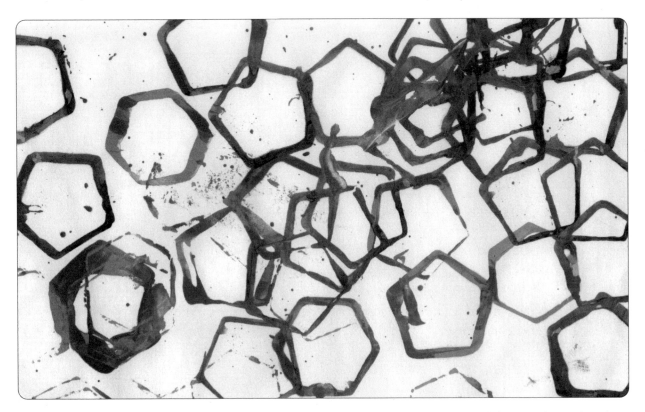

Christopher made this painting with house- and cross-shaped cookie cutters.

"Angry" Stories

Originating ideas

In the last few days there has been a lot of physical conflict in your group during small-group and snack times. Children have been arguing over which chairs to sit in, whom to sit next to, and how much juice their classmates should pour into their glasses. These arguments have resulted in children hitting one another, knocking over chairs, and pushing classmates.

Key experience—Initiative and social relations: *Expressing feelings in words* and *solving problems encountered in play*

Possible additional key experiences

Language and literacy: *Reading in various ways—reading storybooks, signs and symbols, one's own writing*

Materials

✔ The book *Angry Arthur* by Hiawyn Oram

✔ **Markers**

✔ **Blank books,** one per child, made by stapling together two or three sheets of **construction paper** or **white paper**

✔ Back-up materials: **magazines, scissors,** and **glue**

Beginning

Engage children in a conversation about the pushing, hitting, and other problem behaviors that have been going on in the classroom over the past few days. After children have had a chance to talk about their actions and feelings, tell them you've brought a story about a boy who also gets angry, and then much later begins to calm down. Read the book to them.

Middle

As you are reading the book, be sure to ask for children's comments. Pause frequently to acknowledge and discuss their reactions as you read through the book. When you and the children are finished reading and discussing the story, pass out the blank books and markers, and tell children they can use the materials to tell their own angry stories. Watch to see if children use the paper and markers to depict their feelings, discussing what makes them mad or what they do about their own anger or frustration. Encourage children to tell their stories aloud as they show you their books; to "write" their own narration by drawing pictures, scribbling, or using other writing forms; or to dictate words for you to record on their pages. Understand that some children will simply color on the pages without making a connection to their own or other people's feelings.

End

Give a 2-minute warning before starting cleanup. Make the transition to the next activity by asking children to imagine that they were angry, but like Arthur in the story, they have calmed down. Ask them to walk to the next activity as if they were not angry anymore. As children move to the next activity, watch the ways they use movement to express this calmed-down state.

Follow-up

1. Add *Angry Arthur* and some of the child-made books to the book area.

2. Put blank books, notepads, and markers near the book area so children can continue their book-making efforts.

Tapping, Beating, Swaying, and Singing

Originating ideas

Audie, Patrick, and James recently created a work time game in which they use a large, hollow, ramp-shaped wooden block and rhythm sticks. In the game, one boy holds on to the top of the block while running in place on the inclined surface of the block. While one boy is running on the block, the other two boys get rhythm sticks and tap out a beat on the floor, matching the pace of the runner. As they tap, they chant these words: "Run, run, run in place, while we hit the sticks."

Key experience—Movement: *Feeling and expressing steady beat*

Possible additional key experiences

Creative representation: *Imitating actions and sounds*

Language and literacy: *Reading in various ways—reading storybooks, signs and symbols, one's own writing*

Music: *Singing songs*

Time: *Starting and stopping an action on signal*

Materials

✔ A **tape player** and High/Scope's ***Rhythmically Moving 3*** recording or another suitable recording

✔ **Rhythm sticks or rounded chopsticks,** one pair per child

✔ **Paper plates,** two per child

✔ For back-up: **things to drum on—empty oatmeal containers, cookie tins**

Beginning

Sit in a circle with the children. Begin tapping the top of your head in a steady beat, saying the word *head* each time you touch it. As soon as most of the children have begun imitating your actions, switch to another body part, such as the knees. Ask children to suggest one more part of the body that they can tap. Then tell children you will be putting on music for them to tap to, and that they will be able to choose the parts of their own bodies that they will be tapping with and tapping on. Put on the selection "Alley Cat" from the recording and tap with the children, encouraging each child to tap the beat as he or she pleases.

Middle

When the selection ends, tell children you will play the song two more times. Suggest that this time they tap a part of their body (such as their shoulders or knees) with the paper plates they are holding in each hand. When children are finished, pass out the rhythm sticks, and as you play the song again, have them use the sticks to tap the steady beat on the floor. Notice which children are maintaining the steady beat and which are tapping in a random fashion. End this

Noticing that one child is holding the chopsticks like a violin, the teacher imitates his actions and the other children follow suit.

part of the group by asking children to find a partner, sit face to face with that partner, hold hands, and sway back and forth while singing "Row, row, row your boat."

End

With children, put away the paper plates and rhythm sticks. To make the transition to the next activity, play a flying game. The group chants, "Fly, fly, fly your arms. Fly your arms, then stop," while flapping their arms and moving to the location of the next part of the daily routine. The group stops on the word *stop,* then begins chanting, and moving, again. This is repeated until all the children have reached the destination.

Follow-up

1. Bring the rhythm sticks to recall time. As you and the children tap a steady beat with the sticks, make up a chant or song about the children's work time experiences. Incorporate children's suggestions. For example, if Trey played with Play-Doh, you could chant and tap, "Trey played with Play-Doh, yes he did." Repeat two more times, then end with "in the art area."

2. Add rhythm sticks to the music area.

Super Bubbles

Originating ideas

Recently at work time three of the children were talking about the different-sized soap bubbles they were making as they splashed water and dishwashing liquid around in the water table.

Key experience—Seriation: *Comparing attributes (longer/shorter, bigger/smaller)*

Possible additional key experiences

Seriation: *Arranging several things one after another in a series or pattern and describing the relationships (big/bigger/biggest, red/blue/red/blue)*
Movement: *Moving in nonlocomotor ways (nonanchored movement—running, jumping, hopping, skipping, marching, climbing)*
Time: *Experiencing and describing rates of movement*

Materials

✔ **Bubble solution** prepared by mixing 2 cups of Joy or Dawn **dishwashing liquid,** 6 cups of **water,** and 3/4 cup **light corn syrup** (Let the liquid settle for several hours, then stir or shake before using.)
✔ **Shallow containers,** such as pie plates or cookie sheets with sides
✔ **Bubble tubes** made by covering one end of a cardboard tube with a piece of a nylon stocking
✔ **Bubble wands** and **fly swatters**
✔ For back-up: **food coloring**

Beginning

Bring one container of the bubble solution, a bubble tube, and a fly swatter to the table. (Set the rest of the materials outdoors if possible.) Ask one child to dip the bubble tube in the solution and then blow. Listen to children's reactions: Do any of them describe the bubble (the bubble tube should make long, thin bubbles) or the actions that resulted in the bubble? Then have another child dip the fly swatter in the solution and shake the handle. Again, wait for children to describe the bubble shapes and the differences or similarities among the bubble shapes created by the two methods. Notice which children use comparison words. Tell children where the rest of the bubble-making materials are (preferably outdoors where children can observe bubbles being carried by the wind and soaring to greater heights). Encourage children to begin exploring the materials.

Middle

Blow bubbles alongside children. Listen to and comment on their descriptions of the bubbles they are making. Accept children's descriptions of their own bubbles ("I got a short one, then a smaller short one"). Ask children to give you directions for making bubbles like the ones they are making. Suggest new places to make bubbles from (first from the top of a hill, then from down below). Refill the solution as needed, with children's help, commenting on how the solution looks before it turns into bubbles.

End

When small group is over, point to the storage containers and ask children to put the materials away. Encourage them to make bubbles as they walk toward the containers.

A small pail of solution and a fly swatter make lots of bubbles.

Follow-up

1. Bring a bubble wand to planning or recall time. Ask children to blow a bubble toward the area where they plan to work (or worked). If you make this a planning strategy, be prepared for those children who decide to incorporate bubble blowing into their work time plans.

2. Bring bubble-making equipment back outdoors. Add additional bubble-making devices, such as straws, hoops made from stretching nylon over plastic wire, commercial bubble pipes, and plastic rings from six-packs of canned soda.

Clothesline Running

Originating ideas

Several children in the class have been playing a made-up game of running and falling, both during work time and at outside time. The children chase one another around the classroom or yard, then fall in a heap on the ground or floor, tickling one another and laughing.

Key experience—Initiative and social relations: *Building relationships with children and adults*

Possible additional key experiences

Movement: *Moving in locomotor ways (nonanchored movement— running, jumping, hopping, skipping, marching, climbing)*

Time: *Experiencing and describing rates of movement*

Materials

✔ A long length of **clothesline,** hung horizontally about 5 feet from the ground

✔ **Towels and sheets** draped or clipped with clothespins along the length of the clothesline (leave gaps between the cloths)

✔ Back-up materials: **big, soft balls** (for example, beach balls or foam balls)

Beginning

Gather outdoors near the prepared clothesline. Tell children you have noticed their interest in running games and that you thought it would be fun to run through the cloths hanging on the line. For safety reasons, establish a running direction.

Middle

Run with the children, pointing out how the sheets and towels get pushed up, then drop down again as you run past them. Encourage children to run in pairs, asking a child to be your running partner for one run through the clothes. Watch children's movements as they run. Are they coordinated and sure-footed, or do they stumble and hesitate as they get to the clothing on the line. Notice whether children make up group games, like chasing each other or aiming for the open spaces between the sheets and towels as they run.

End

When small-group time is nearly over, tell the children they have time for two more runs. Afterwards, take down the sheets and towels, and enlist children's help in folding the items and storing them in a laundry basket nearby.

Follow-up

1. Bring one of the sheets to large-group time and have children gather around it. As you play both fast and slow musical selections, have the children lift the sheet up and down in time to the music.

2. Cover the planning table with the sheet and have children sit under the table to describe their work time plans to one another.

As a variation on the small-group experience described here, teachers in this Portuguese preschool planned a small-group time in which children worked with clothespins, wet clothes, and a clothesline.

Swinging at Targets

Originating ideas

Some children have been talking about the way their siblings "bat" at balls, then run the bases during T-ball games.

Key experience—Language and literacy: *Talking with others about personally meaningful experiences*

Possible additional key experiences

Movement: *Moving in locomotor ways (nonanchored movement— running, jumping, hopping, skipping, marching, climbing)* and *moving in nonlocomotor ways (anchored movement—bending, twisting, rocking, swinging one's arms)*

Materials

✔ Various kinds of **real or makeshift bats** (large plastic and foam baseball bats, lengths of plastic pipe, wooden dowel rods, large Tinkertoy pieces, child-sized plastic rakes, shovels, or brooms)

✔ **Materials to bat at,** including various **balls** (beach balls work well because they are large and lightweight) and **hanging materials** suspended from poles, tree limbs, basketball hoops, or other structures (for example, plastic grocery sacks filled with wadded up newspaper or plastic milk containers)

✔ For back-up: **flat plastic objects** to use as bases (for example, lids from plastic containers)

Beginning

Meet with children in a space large enough to allow them to safely run and swing at their targets (outdoors or in a gym). Point out the prepared targets hanging from tree limbs, poles, or other structures. Tell children you have also brought materials for batting at the targets. Discuss safety considerations, asking children what they think would happen if they hit another child with a bat. Encourage children to make some rules for playing safely.

Middle

Play alongside children, swinging at the targets with your own batting tool, acting as the pitcher who throws objects for children to aim at, hitting the balls children throw to you. As you interact with children, watch the ways children use their bodies to run and swing. Do they swing with a sideways motion at the targets or do they hit at them with up-and-down movements? Is their running sure-footed and well coordinated, even when they encounter obstacles such as other balls on the ground or tree stumps? Do they follow rules as they attempt to play with others (hitting the ball, then running) or are they playing by themselves, exploring the materials or creating solitary games?

End

With the children, put materials away in their regular storage containers, one for the batting tools, the other for the target objects.

Follow-up

Make a map to use at planning or recall times. On a large sheet of paper draw a baseball diamond and put symbols for the interest areas along the baseline. Give children, one at a time, a teddy bear counter or small person and ask them to "run" their prop to the area of their work time choice.

Animal Sounds and Movements

Originating ideas

Since returning from a field trip to a 4-H youth livestock show (see p. 224), children have been imitating the sounds they heard animals make at the livestock show.

Key experience—Music: *Exploring and identifying sounds*

Possible additional key experiences

Movement: *Expressing creativity in movement*

Materials

✔ **Taped musical selections**

✔ Six **pictures of different animals** seen at the fair (you can use drawings, actual photos, or pictures cut out of books or magazines) and a **cube** about 6 inches across (tape a different animal picture to each surface of the cube.)

✔ A **tape player**

✔ For back-up: **photos** taken on field trip and/or **books** about farm animals

Beginning

Gather in a large indoor or outdoor space with children and have the group form a circle. Show them the animal cube and tell them that the group will be using it to play a music and movement game. Ask a child to throw the cube into the middle of the circle. When the cube lands, explain that the group will be pretending to be the animal that is face up on the cube and that everyone will be imitating the movements and sounds of the animal in their own ways. Then everybody imitates the animal, moving freely within the space.

Middle

Have the group form a circle again. Explain that you will be continuing the same game, but that this time you will be playing music. Ask a child to throw a cube, and have the group imitate the animal shown as you play the music. When you stop the music, have everyone

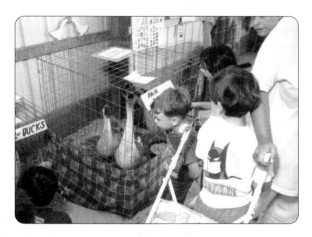

Children's visit to a livestock show is an opportunity to experience the sounds, sights, textures, and smells of a variety of animals. The day after the field trip, this small-group time gave children a setting for representing their experiences at the show.

stop and ask another child to throw the cube again to choose another animal to imitate. Each time you stop to choose a new animal, pause to discuss the children's ideas. Encourage children to describe the animal movements and sounds they will be creating. Pose questions to the group, such as "How would this animal sound if there were food in its mouth?" "What if her owner were gently stroking her fur?" "What would the horse look like if it were in a hurry to get back to the stable?" After putting on the musical selection, join with the children as they carry out their movement and sound ideas. Repeat this game until children lose interest.

End

Tell children there is time for one more cube toss. Play the last musical selection, then ask children to pick their favorite animal to imitate as they move to the area for the next part of their daily routine.

Follow-up

1. Bring the animal cube to the next day's planning time. Have the children throw the cube into the area they plan to work in, noticing which animal is on top when the cube lands. Then have the child move like that animal to the chosen area and then discuss his or her plans.

2. Put the animal cube and the taped musical selections you used in the area of the room where the tape player is stored for children's use at work time.

Stone Drawings

Originating ideas

At outside time children have been digging up stones, putting them in buckets, then lining them up on the flat surface of the pavement.

Key experience—Creative representation: *Drawing and painting*

Possible additional key experiences

Language and literacy: *Writing in various ways—drawing, scribbling, letterlike forms, invented spelling, conventional forms*

Initiative and social relations: *Building relationships with children and adults*

Classification: *Distinguishing and describing shapes*

Seriation: *Comparing attributes (longer/shorter, bigger/smaller)*

Materials

✔ A variety of different-sized **stones and rocks**

✔ A long piece of **white butcher's paper** on a long tabletop or other flat surface

✔ For back-up: **glue**

Beginning

Gather around the paper or paper-covered table with children. Tell them, "Today I brought something you can use to draw lines, shapes, and designs on the paper; but, it's NOT Magic Markers." Show them the stones. Explain to children that they can arrange the stones to make the designs or pictures they want—in this way they will be "drawing" without actually using pencils, markers, or crayons.

Middle

Look for the ways children arrange and rearrange the stones on the paper. Do they make straight lines, lines with curves, actual shapes? Are they working alone, or do they pair up with others to make a single design? Are they comparing the stones, or arranging them in patterns according to their sizes, weights, or any other characteristics? Comment on children's work, if you can do so without interrupting them. Describe what you see—lines, shapes, textures, and so forth—rather than make assumptions about what a stone picture might represent.

End

If possible, leave the stones and the paper or table in place for several days so children can continue to explore the various ways they can arrange stones. If this is not possible, store stones in a container and ask children for their ideas on where these would best fit into the classroom interest areas.

Follow up

1. Repeat this activity, this time having children arrange colored stones (painted by children) on a contrasting surface, either paper or a tabletop in another color or texture.

2. Add stones to the classroom area or areas as suggested by children.

3. Prepare a long sheet of paper with the symbols for the interest areas drawn across the top. At planning or recall time, ask children to place a stone on the spot representing each area they plan to work in (or did work in). Before talking about their plans or experiences, comment on the designs the stones make and relate this to the numbers of stones placed on each area symbol.

Feeling Dance

Originating ideas

At outside time children have been playing a made-up game in which they pretend to be various kinds of monsters as they chase one another. When one of the "monsters" catches another child, the captor acts out the behavior or emotion associated with his or her character (the "laughing monster" laughs at his captive, the "scary monster" makes roaring sounds, the "angry monster" shakes his finger at the captive and knocks him or her over).

Key experience—Movement: *Expressing creativity in movement*

Possible additional key experiences

Creative representation: *Imitating actions and sounds*

Music: *Moving to music*

Time: *Starting and stopping an action on signal*

Materials

- ✔ A variety of recorded **musical selections** in various styles, for example, classical, jazz, reggae, and pop recordings
- ✔ A **tape player**
- ✔ **Carpet squares** (one for each child) scattered throughout the space
- ✔ For back-up: a selection of **books** that lend themselves to a discussion of feelings, such as *Isabella's Bed* by Alison Lester, or *Sadie and the Snowman* by Allen Morgan

Beginning

Gather in a large, open space. Begin singing the familiar song "If You're Happy and You Know It." Instead of singing about clapping hands, stamping feet, and so forth, ask children for their own ideas for motions or words that might express happiness. Put their words or actions into the verses. After a few turns with the word *happy*, switch to another feeling word, such as *scared* or *angry*.

Middle

When the singing is over, tell children you will put some music on for them to listen to for a short while. After you turn the music off, ask them how the music made them feel. After they listened to a selection and identified an emotion associated with it (for example, *happy, sad, angry),* ask them to show that feeling by moving their whole bodies. Put on the same music and participate with them as they make their happy, sad, or angry motions. When the music stops, have them jump to a carpet square to ready themselves for the next musical selection. Then follow the same process for the next musical selection (and emotion).

End

Tell children you will be playing one more musical selection. Once children have moved their bodies to the last song and are standing on carpet squares, have them put away the squares in a big pile while pretending they are tired from all the moving.

Follow-up

1. Leave the musical tapes in the block area for future work time listening and dancing.

2. The day after your small-group activity, play a stop-start musical game with the children at planning or recall times. Tell children that when you turn on the music, they should go get something they plan to work with or did work with. When the music stops, everyone freezes until it begins again. Play this game until everyone has returned to the planning/recall table with an item from the classroom.

Decorating Tube and Cup Sculptures

Originating ideas

This activity is designed to follow up on the "Tube and Cup Sculpture" activity (p. 56) conducted on the previous day. During that activity, the teacher observed children making up imaginative stories about their creations, which included such descriptions as "binoculars to look through with your eyes," "a city with tall buildings," and a "set of steps to climb you up to the moon." This small-group time was planned to encourage children to build on these stories.

Key experience—Language and literacy: *Describing objects, events, and relations*

Possible additional key experiences

Creative representation: *Relating models, pictures, and photographs to real places and things*

Materials

✔ **Children's sculptures** from the day before

✔ **Materials for decorating (glue, paint, paintbrushes, buttons, pipe cleaners)**

✔ **Newspapers** covering the table

✔ **Smocks** to protect children's clothing

✔ A bucket of **soapy water** with **sponges** for cleaning the table top

✔ For back-up: **markers**

Beginning

Bring children's sculptures to the table. Put the materials for decorating on the table, and tell children you've brought some things they can use with their sculptures.

Middle

Set out the decorating materials. You may want to start with just the paint, adding some or all of the other materials if children's interest in painting begins to wane. Watch to see whether children choose decorating materials related to the stories they made up about their work the day before. For instance, the child who described his creation as "a city with tall buildings" might attach a pipe cleaner to the top of his cardboard tube and call it a smokestack. As children add materials, listen for any spatial language they use to describe their efforts ("I'm gonna' paint the whole tube, inside and out").

End

When children have completed their work and have had a chance to describe the decorations they've added, ask them to put away their sculptures in the spot where you leave artwork to dry. Then have children return to the table to clean up.

Follow-up

1. Display children's sculptures throughout the room at their eye level. If children have described their efforts, their words may be written down to become part of the display.

2. Add decorating and sculpture-making materials to the art area for use in future work time creations.

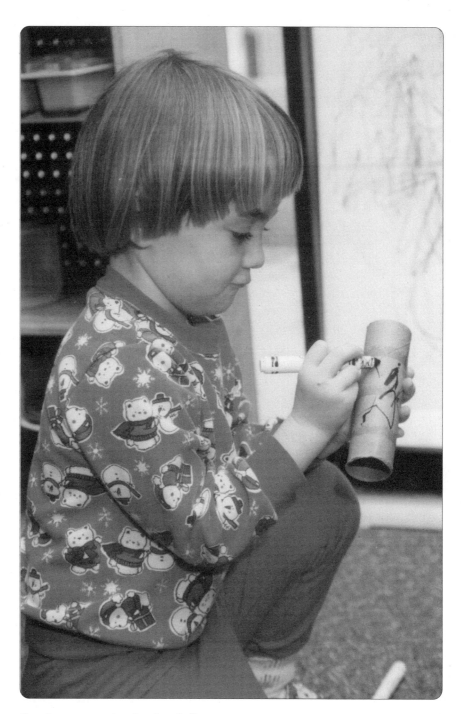

Drawing on a round surface is a challenge.

Texture Collage

Originating ideas

The other day one of the children came to school with an accordion-shaped piece of plastic packing material that had come with a present his grandparents sent him. Children gathered around him at greeting circle, taking turns running their hands and fingers over the ridges in the plastic.

Key experience—Classification: *Exploring and describing similarities, differences, and the attributes of things*

Possible additional key experiences

Creative representation: *Making models out of clay, blocks, and other materials*

Seriation: *Comparing attributes (longer/shorter, bigger/smaller)*

Materials

✔ Scraps of **textured materials (foam, carpet, cotton, yarn, coarse and fine sandpaper, textured wallpaper samples, felt pieces)** in a variety of shapes and sizes

✔ **Glue** and **glue brushes**

✔ **Scissors**

✔ **Heavy cardboard** to act as a base for each child's creation.

✔ A **shoe box** (lid on) filled with samples of the textured scraps, with a hole in the side large enough for a child's hand to fit through

✔ For back-up: **glitter**

Beginning

Bring the shoebox to the table and encourage children, one by one, to stick one of their hands through the hole, feel and describe an item that is inside, and then pull it out of the box for all to see. When several children have had a turn at this, give each child a piece of cardboard, a bottle of glue, a glue brush, and a container holding scraps of textured materials.

Middle

Observe the different ways children explore the materials. Some children will take scraps of material and touch them with their fingertips, or rub them against their faces. Others will begin gluing items to their cardboard in a random way. Still others will use the pieces very purposefully, cutting scraps to reshape them into specific representations (a house with a chimney). Observe which aspects of this activity are most interesting to the children (exploring different textures, combining materials in different ways, using a brush to spread the glue) and comment on those interests.

End

With the children, collect the scrap materials in a large box to be stored in the art area. Encourage children to share their thoughts about their work as they are cleaning up. Make comments that are relevant to individual children's work ("I noticed your cardboard only has soft things on it"), and note whether children expand on your observations.

Follow-up

1. Add the box of scraps to the art area.
2. Repeat this activity, this time having children glue their scrap materials to three-dimensional items (a box, a ball, tubes).

Paper-Plate Streamer Dancing

Originating ideas

This activity is a follow-up to the "Paper-Plate Decorating" activity (see p. 96). After the activity, children took their paper plates outside and moved with them in a variety of ways, for example, holding them high above their heads while running down a hill, swinging them up and down and side to side while standing still, and twirling around and around while holding them straight out from the sides of their bodies.

Key experience—Music: *Moving to music*

Possible additional key experiences

Movement: *Moving in nonlocomotor ways (anchored movement—bending, twisting, rocking, swinging one's arms); moving in locomotor ways (nonanchored movement—running, jumping, hopping, skipping, marching climbing);* and *expressing creativity in movement*

Time: *Starting and stopping an action on signal*

Materials

✔ Small-sized **paper plates,** one per child, with **streamers** (cloth, crepe paper, or ribbon) of various lengths tied to one side of each plate
✔ Recorded **musical selections** in at least three different styles (for example, salsa, classical, and jazz)
✔ A **tape player**
✔ For back-up: a basket of **musical instruments**

Beginning

Meet in a large, open area. Tell children you've brought three different kinds of music for them to move to. As you play an excerpt from each recording, encourage children to move to the music while seated on the floor. Observe the different ways the children move their bodies. Encourage children to talk about the ways the music makes them feel like moving and the reasons why.

Middle

Pass out the paper plates with streamers tied to them, and tell the children that you will play the same musical selections as before but that this time they can stand and move around while using the paper plates. When the music stops between selections, have children freeze and again encourage them to describe their actions. Notice the ways children move to the music. Do they do so from sitting or standing positions? Are their movements in time to the beat of the music? Do they engage in movements alone, or with other children?

End

Tell children there is one more dance selection left to play. When the selection is over and children have frozen in their places, place a basket for the paper plates in the center of the floor. Tell children you are putting on a SHORT musical selection so you can see how many times it will play before they are all finished putting their paper plates away and sitting back down on the floor.

Follow-up

1. Store the paper plates near the tape player and taped music so children can choose these materials during future work times.

2. Play some of the musical selections used in this activity during cleanup time. As a strategy for encouraging children to clean up quickly, count the number of times you can play a selection until the room is completely cleaned up.

3. Take the children on a field trip to a local dance or karate studio where they can see others moving their bodies in unusual ways.

IV

Small-Group-Time Plans Originating From

Community Experiences

Neighborhood Treasure Walk

Originating ideas

The center is located in a neighborhood that includes gardens, a fruit and vegetable market, a bakery, street musicians, a church with bells, a fire station, seasonal lawn ornaments and window displays, and a construction site. From time to time throughout the year, teachers take children on a walk around the block so children can observe and sometimes participate in neighborhood events and changes.

Possible key experiences

Creative representation: *Relating models, pictures, and photographs to real places and things*

Language and literacy: *Reading in various ways—reading storybooks, signs and symbols, one's own writing*

Classification: *Exploring and describing similarities, differences, and the attributes of things*

Space: *Experiencing and describing positions, directions, and distances in the play space, building, and neighborhood*

Materials

✔ A **chart** that describes in pictures and symbols which adult and which children will be in each walking group (Note: For this small-group time, you will need to have enough adults so that the adult-child ratio is 1 to 4 or 5, so you may need to invite parents and other volunteers to accompany you on the walk.)

✔ A **points of interest guide** for each adult, describing things along the walk that may be of interest to the children, such as railroad tracks, someone's vegetable garden, a huge maple tree with a tree house built inside (This will encourage the adult with each group to give children enough time to pause and freely explore things they encounter on the walk.)

✔ A **camera** to photograph the experiences (An instant camera is preferable so there is no delay time.)

✔ **Bags or backpacks** for collecting found objects, one per walking group (to be carried by the adult, so that children have their hands free to explore)

✔ For back-up: for each child, a simple **snack** (such as a granola bar and, if the day is hot, a juice box)

Beginning

Gather in your regular small-group time meeting place. Explain that the group will be taking a neighborhood walk for your small-group activity. Discuss the chart with children, encouraging them to use it to figure out which adult and which children will be in their walking group. Give each adult the list of children they are responsible for and the points of interest guide. Collect the camera, pass out the backpacks, put on adequate clothing, and set out.

Middle

As you walk, talk with children about things of interest to you both. Listen for the observations and comments they make (for example, in the winter, they may notice tire treads in the snow). Encourage children to collect any "treasures" they find. (A treasure can be anything they consider valuable—acorns with tops, bottle tops, sticks, unusual stones.) Take photographs of items of interest (a pumpkin flag hanging on someone's front porch).

End

When you return to the classroom, have each adult empty the contents of the bags or backpacks. Look together at the photographs taken on the walk. Encourage children to discuss what they saw and did on the trip. Some of the children, who have taken previous walks in the same neighborhood at other times in the year, may discuss changes they've noticed in the neighborhood.

In a follow-up small-group time, children decorate leaves collected on the treasure walk, combining materials in unique ways.

Follow-up

1. Use the materials you collected in a small-group time the next day (see "Working With Treasure Collections," next).

2. If appropriate, add the materials collected to the interest areas.

3. Display the photos taken at children's eye level, or transfer them to a photo album that is located in the book area. Have conversations with children that encourage comparisons of new photos to those taken on walks around the neighborhood earlier in the year.

Working With Treasure Collections

Originating ideas

This small-group experience is a follow-up to the previous one, in which children periodically repeat a walking route in the neighborhood of the school or center so they can experience neighborhood events and changes along the route. During the previous day's walk, adults carried backpacks and small containers so children could collect "treasures."

Possible key experiences

Classification: *Sorting and matching*

Language and literacy: *Talking with others about personally meaningful experiences*

Seriation: *Comparing attributes (longer/shorter, bigger/smaller)*

Materials

✔ **Materials collected by children,** which will vary according to the season and what is available in the school environment (examples: pine cones, leaves, horse chestnuts, acorns, sticks, stones, rocks, dirt, empty candy wrappers, and other litter)

✔ **Magnifying glasses,** one per child

✔ For back-up: a selection of **art area materials**—for example, **paint, paintbrushes, glue, glitter,** or **paper**—that children can use to represent some aspect of the walk or to decorate the "treasures" they collect

Beginning

Spread out on the table the materials collected on the previous day's walk. Say "Here is the collection of things you put in the backpack on our walk yesterday." Watch and listen as children begin exploring their found materials. After a few minutes, add the magnifying glasses to the center of the table.

Middle

Watch the nonverbal and verbal ways in which children choose materials for their own personal explorations. Observe the ways they arrange their materials and use their magnifying glasses, and listen to the words they use to describe their actions. Look for the ways children combine materials from the walk. Listen for any conversation identifying specific places where particular materials were collected. Follow up on what you observe by describing and labeling children's efforts ("I see you put only pine cones in this pile") and building on their interests ("We did find these under the oak tree. I wonder if they'll be there again next time we take a walk").

End

Ask for children's ideas on where to put any leftover objects from the walk. With children, put away art materials. Have children sort materials they will take home from those to be left at school. Store leftover materials and magnifying glasses in an appropriate spot in the classroom.

Follow-up

1. The day after your walk, use a leftover item as a planning prop. For instance, a child holds a horse chestnut or acorn while sharing work time plans, then rolls the item to another child to indicate who will plan next.

2. Add leftover items to the various areas of the room children have suggested, labeling the container and the storage spot.

Pitching, Throwing, Dunking, Kicking

Originating ideas

Each year the town holds a 3-on-3 basketball tournament to raise money for charity. Several of the children in the classroom have attended the weekend event, in which older children, teenagers, and adults from surrounding communities participated in the competition.

Possible key experiences

Creative representation: *Imitating actions and sounds*

Language and literacy: *Describing objects, events, and relations*

Movement: *Moving in nonlocomotor ways (anchored movement—bending, twisting, rocking, swinging one's arms)* and *moving with objects*

Materials

✔ A variety of **safe objects for throwing** (cloth balls, foam balls, rubber balls, wadded-up newspapers, small basketballs)

✔ **Containers** children can aim for located in different spots and at different height levels (Some can be placed directly on the ground; others, on steps or ledges; still others, attached to fence posts or other high objects.)

✔ For back-up: **Props** for children who wish to pretend to be members of the audience or announcers—**blankets, stuffed animals, chairs, microphones**

Beginning

Ask children who were at the basketball tournament to describe what happened for the group. See if one or several children would be willing to act out some of the motions they are describing. Tell the children that today you brought some different kinds of balls for them to throw into containers; then take them outside to show them the balls and "baskets."

Middle

Pick up a ball and begin bouncing it and then throwing it into one or more of the containers. Observe the kinds of balls that children choose, the containers they choose to shoot for, and whether they play by themselves or with others. Don't be surprised if children begin some throwing games with one another, or kick the balls instead of throwing them, or simply run with balls in their hands. Participate in their play, imitating the things you see them doing and the sounds you hear them making.

End

When there are only a few more minutes left in small-group time, announce this to children. Ask them to describe their favorite ways of using the balls and containers or to talk about which ball they liked the best. Put away the balls in a large container and gather smaller containers together in one spot.

Follow-up

1. For planning time the next day, set up baskets with area symbol cards near or on them. Ask each child to find an item in the classroom he or she plans to use, to toss it into the basket representing the interest area the item came from, and then to describe his or her work time plans.

2. Bring the containers back outside the next day, and ask children to find places to put them to create their own ball games.

Children's interest in playing basketball was still high after this small-group time, so teachers added child-sized basketball equipment to the playground.

Flannel Board Storytelling

Originating ideas

The father of one of the children in the preschool is well known in the community for his storytelling skills. Recently he visited the classroom and told a story using the flannel board, encouraging children to participate by asking them to act out different aspects of the story.

Possible key experiences

Creative representation: *Relating models, pictures, and photographs to real places and things* and *pretending and role playing*

Language and literacy: *Having fun with language—listening to stories and poems, making up stories and rhymes*

Materials

✔ Individual **flannel boards** for each child, easily made by stapling a piece of flannel to a piece of sturdy cardboard

✔ **Flannel pieces** representing people, their clothing, trees, the sun, different animals, and so forth (These may be homemade or purchased.)

✔ For back-up: plain **paper** and **markers**

Beginning

Start by telling a story about a storyteller who came to the school and told all the children a story. Include some of the highlights of the real storyteller's story in the one you re-create. Pause frequently to give children the opportunity to comment about their own recollections of the storyteller. Keep your story short—its purpose is just to launch children in the direction of creating their own stories. When you are finished with the story, tell children you've brought materials for each of them to use in making up their own stories.

Middle

Watch as children manipulate the flannel pieces. Some may lay pieces on top of one another; others may lay them side by side; others may sort or match pieces in various ways (for example, pairing flannel animals with flannel people). If children tell stories, listen carefully as they tell them. Notice which children use simple sentences about individual flannel pieces and whether any children make up well-developed stories that include many details. Observe whether any of the children try to engage others in their storytelling by asking them to listen to or to act out the story using the flannel pieces. Tailor your interaction strategies to the individual child activities you observe. For example, you could support individual children by arranging and rearranging flannel pieces, repeating children's descriptions of flannel pieces, or taking on a story role suggested by the storyteller.

End

As a transition to the cleanup phase of this activity, make up a story about a small-group time in which children played with flannel board pieces and then put them away. See if any children add to the story details that relate to cleaning up and storing the materials.

Follow-up

1. Add storybooks to your classroom that have pictures only, no words. One example is *The Snowman* by Raymond Briggs.
2. Put the flannel boards and the flannel board pieces near the book area.

Leaf Blowing

Originating ideas

Each autumn the maintenance person at the preschool uses the leaf blower to make a big pile of fallen leaves that the children run and jump through at outside time. This year several children stood by the window of the preschool watching as he blew the leaves into a pile. They were fascinated by the sounds of the machine and the motions of the leaves as they floated in the air.

Possible key experiences

Creative representation: *Imitating actions and sounds*
Movement: *Describing movement*

Materials

✔ A collection of **autumn leaves** for each child
✔ Materials to blow through **(plastic piping, toilet paper rolls, straws)**
✔ For back-up: paper **fans**

Beginning

Gather around a table that you have positioned near or in a large open space, and put some leaves in the middle of it. Tell children that you noticed them watching the person blowing leaves with the machine on the previous day and that you've brought some materials for them to use that will make the leaves on the table blow around. Set out the straws, pipes, and/or toilet paper rolls, and begin using one of them to move the leaves across the table.

Middle

Look for the kinds of choices children make. What are they trying to do with the leaves? Do they blow them up into the air? Do they let them drop on the ground or table before they blow at them again, or do they try to keep them up in the air? Based on these observations, engage yourself with the children: for example, imitate the ways they are blowing their leaves, describe the places your own leaves are blowing toward, or try playfully starting a game of blowing leaves back and forth between you and another child or children. Listen to the words children use to describe their actions, and watch for any cooperative games children may devise.

End

Before gathering up the leaves and putting the tubes away, ask children to describe one place their leaves went or what they used to move the leaves around.

Follow-up

1. Take the blowers outdoors so children can continue their play in a larger area, with wind as a factor.

2. For planning or recall the next day, have children blow a leaf into the area they plan to work in (or did work in), then describe their work time plans or experiences for the others.

3. Add blowers and leaves to one of the classroom areas, and put additional materials (such as feathers or cork disks) nearby so children can compare the effects of blowing on different kinds of objects and materials.

Pumpkin Faces

Originating ideas

Several days before this small-group time was planned, the preschool class took their annual field trip to a local pumpkin patch. When children arrived at the farm, the farmer seated the children in a hay wagon attached to a tractor, then drove the children to the pumpkin patch. On the way to the pumpkin field, they passed a display of large pumpkins with painted faces. The farmer stopped for a moment next to each pumpkin, so children had the opportunity to look closely and comment. When they reached the pumpkin patch, each child chose one small pumpkin from those lying in the field. They returned with their pumpkins and stored them in the classroom.

Possible key experiences

Creative representation: *Drawing and painting*

Language and literacy: *Describing objects, events, and relations*

Initiative and social relations: *Making and expressing choices, plans, and decisions*

Materials

✔ A small **pumpkin** for each child, preferably one selected by the child

✔ **Markers** (Test first to see if children can use them on pumpkins; some kinds of markers will rub off, leaving children frustrated.)

✔ **Stickers** (for example, magazine stamps or stickers depicting seasonal symbols)

✔ **Photographs** of the field trip

✔ For back-up: additional **decorating materials—glue and glue brushes, hay, straw, yarn, glitter**

Beginning

Pass around the photos taken on the field trip. Give children time to recollect the experience and discuss things they remember. Encourage children to comment on the things they see in the pictures. When conversation slows down, pass out the pumpkins, set out the rest of the materials, and tell children you've brought materials they can use to make their own pumpkin faces.

During a field trip to a pumpkin patch, each child chooses a pumpkin. The next day at small-group time, children will decorate their pumpkins.

Middle

Watch as children select materials for decorating their pumpkins. Expect that some children will cover their pumpkins with patches of color, and others will enjoy using their pumpkins as unique, three-dimensional drawing surfaces. Children may draw all kinds of things on their pumpkins, including both abstract designs and actual representations of facial features. Some children may prefer not to draw at all and may instead cover their pumpkins with stickers.

End

When there are just a few minutes left in small-group time, tell children they have time to add one or two more things to their pumpkins. Find a place for children to display their pumpkins. Observe the ways children carry their pumpkins to the designated spot, noticing which children's movements are coordinated (carrying pumpkins without bumping into things) or complex (carrying pumpkins while walking backwards). Encourage children to participate in putting away all other materials and washing the table with warm, soapy water.

Follow-up

1. Bring the decorated pumpkins to large-group time. Put on some lively dance music and have the children do a dance around their pumpkins. Each time the music stops, have them stop dancing and stand near their pumpkin.
2. Make a flannel board story with pumpkin faces, and add it to the book area.

Painting Leaves and Plants

Originating ideas

This activity is another follow-up to the "Neighborhood Treasure Walk" activity, p. 176. The plan for this activity was developed after a walk in which children had collected a variety of leaves and plants. Conversation had focused on the different shapes and colors of the leaves and plants children were observing.

Possible key experiences

Creative representation: *Drawing and painting*

Language and literacy: *Talking with others about personally meaningful experiences*

Classification: *Sorting and matching* and *using and describing something in several ways*

The teacher moves around the group, observing and imitating what children are doing.

Materials

✔ A selection of **leaves** of different shapes (for example, willows, palm, eucalyptus, oak, ginkgo, maple, chestnut); **grasses;** or **plants** (use whatever grows in your area)

✔ **Styrofoam meat trays, cardboard pieces,** or **paper plates,** one per child, to use as trays to hold the leaves

✔ **Paints** and **painting smocks**

✔ For back-up: **glue**

Beginning

Remind children of their walk the previous day, and listen to their recollections of the event. Tell them you've brought some of the leaves, grasses, or plants they collected, along with some paint. Say that you remember how fascinated they were by the colors (or shapes) of the materials they picked up on the walk. Give each child a meat tray or cardboard piece labeled with his or her name and a selection of the leaves, grasses, or plants, several of each variety.

Middle

Notice the different ways children use materials available to them. Some children may make a game out of grouping the grasses, plants, or leaves. To gain information on children's classification abilities, observe the ways these children group materials: look for groupings of identical or similar materials and listen for children's descriptions of their efforts. Other children may arrange their materials randomly on their trays. Some children may choose not to use the leaves, grasses, or plants at all, and may instead decide to paint their trays. Observe how they paint—for example, do they cover whole surfaces with single colors or do they add representational or other details to their work?

End

Involve children in gathering the unused materials in the middle of the table, setting aside their projects to dry, and cleaning the paint off the table. Encourage children to talk about their work as they clean up by asking questions or making comments as you clean. ("This goes in the leaf pile—I wonder if anyone used this kind of leaf in their work?" "It looks like someone used yellow paint right here in this spot.") Notice which children respond to the questions and comments.

Follow-up

1. Make a leaf-, grass-, or plant-matching game by pasting some of the plant materials you gathered on cardboard (one item per card), then laminating the cards. Be sure you have additional pieces of the same materials available, or a set of matching cards. The next day at planning time, put a leaf or other piece of plant material (or a card containing the material) on the table for each child and set the deck of cards with the matching items in the middle of the table. Ask a child to pick a card. The child who has a matching card is first to share his or her plans.

2. Store the newly made card game in the toy area.

Scarecrow Stuffing

Originating ideas

As part of the autumn harvest festival, a scarecrow display decorates the front of the elementary school building in which the preschool is located. As the children in the classroom pass by the scarecrows, the teachers have observed children touching the scarecrows' legs, tugging on their arms, and occasionally saying things like "Are you real? Will you dance with me?"

Possible key experiences

Creative representation: *Relating models, pictures, and photographs to real places and things*

Language and literacy: *Having fun with language—listening to stories and poems, making up stories and rhymes*

Initiative and social relations: *Creating and experiencing collaborative play*

Seriation: *Arranging several things one after another in a series or pattern and describing the relationships (big/bigger/biggest, red/blue/red/blue)*

Materials

✔ Three or four different-sized **flannel shirts** and three or four pairs of **blue jeans** in corresponding sizes

✔ **Straw** (You may use straw from the bales described in the "Autumn Straw Jump" activity, p. 194.)

✔ **Newspapers**

✔ **Objects (chairs, hay bales,** and so forth) for propping up the stuffed clothing

✔ The book ***Scarebird*** by Sid Fleischman

✔ For back-up: **paper bags**

Beginning

Read *Scarebird* to the children, pausing often for their comments and encouraging them to compare the scarecrows in the story with those they see in front of the building when they come to preschool. When you are finished reading the story, ask children how they might make their own scarecrows. Show them the materials.

Middle

Observe the ways the children stuff the clothing with the newspaper, noticing whether they work alone or enlist others to help them. Listen for any comments about the ways the clothing changes shape when stuffing is added. Have conversations with the scarecrow you are stuffing in imitation of the comments you have heard them make when passing the scarecrow display (for example, "When you're ready we can dance together"). Notice whether any children make comments about pairing up the jeans with the correspondingly sized shirts.

End

Request children's ideas for ways to accessorize the scarecrows and where to put them. Help them follow through on their additional ideas.

Follow-up

1. Add the book *Scarebird* to your book area.

2. Gather around the scarecrows at planning or recall time and have children tell about their work time plans or experiences.

3. At large-group time, bring out music for dancing around the scarecrows.

Exploring Pumpkins

Originating ideas

This activity is another follow-up to the class field trip to a local pumpkin patch (see "Pumpkin Faces," p. 186). The teachers have reserved four large pumpkins purchased on this trip for use by the two small groups.

Possible key experiences

Language and literacy: *Describing objects, events, and relations*

Classification: *Exploring and describing similarities, differences, and the attributes of things*

Seriation: *Comparing attributes (longer/shorter, bigger/smaller)*

Number: *Counting objects*

Materials

✔ Two large **pumpkins** enclosed in **paper bags** (Beforehand, cut nicely fitting lids in the pumpkins, leaving the pulp and seeds inside—be sure the pumpkin tops are different shapes, to encourage children to figure out which top belongs to which pumpkin.)

✔ A large enclosed table such as an empty **sand and water table**

✔ **Newspaper** or plastic to cover the floor below the table

✔ **Smocks**

✔ For back-up: **magnifying glasses**

Beginning

Gather around the sand and water table (into which you have placed the two paper bags holding the large pumpkins). Ask children to stick their hands inside one of the bags and describe what they feel. Encourage them to guess what is inside and make a game out of their answers ("You think it's a pumpkin and NOT an elephant?"). When the guessing is over, pull out the pumpkins and tell children that for small-group time you'll be opening up the pumpkin to see if it feels the same inside.

Middle

Ask the children to pull off each pumpkin top, then listen for their comments and watch their reactions as they explore the inside texture. Provide some language when they seem interested. ("You're rubbing the pulp between your fingers. It sure feels slimy.") Pick up on children's cues ("I notice you're lining up the seeds. Here's a big clump.")

End

Bring some bowls to the table as small-group time draws to an end. Ask children to put the pulp and seeds in separate containers as they clean up.

Follow-up

1. Bring some pumpkin seeds and interest-area symbol cards to the next day's planning or recall time. Before describing their plans or experiences, have children place a seed on the area symbol card representing their work time choice.

2. Make up a song at large-group time about scooping out a pumpkin. Include the children's descriptions of the experience as the verses of the song.

3. Have potting soil, paper cups, and pumpkin seeds available so each child can plant an individual set of pumpkin seeds. Store containers in a low, accessible spot and have children water their plants. In about a week the seeds will sprout.

Autumn Straw Jump

Originating ideas

Each year the community's elementary school holds an autumn harvest festival, which many of the children in the classroom have attended. The activities at the festival include pumpkin carving, straw jumps, a tricycle obstacle course, and bubble-making. The grounds are decorated with bushel-baskets of apples and scarecrows with pumpkin heads.

Possible key experiences

Language and literacy: *Describing objects, events, and relations*

Movement: *Moving in locomotor ways (nonanchored movement—running, jumping, hopping, skipping, marching, climbing)*

Classification: *Exploring and describing similarities, differences, and the attributes of things*

Space: *Observing people, places, and things from different spatial viewpoints*

Materials

✔ Several **bales of straw** spaced far apart in the outdoor play area

✔ **Bubble-making materials**—several containers of **bubble solution** and a variety of **bubble-makers,** such as **wands, fly swatters,** or **tubes**

✔ For back-up: **tricycles**

Beginning

Bring some bubble-making wands and some straw to the table. Ask the children who attended the autumn harvest festival if they can tell the others about the festival and what they did there. After they have discussed the activities, explain that outdoors you have some materials that they can use for their own school festival. Explain that there are straw bales that they can climb on and jump from, as well as bubble-making materials.

Middle

Run and jump from straw bale to straw bale with the children. Blow bubbles from various places (sitting or standing on top of a bale, or crouching down low next to one). Comment on the things you can see when standing on a bale, as opposed to being on the ground.

Wonder aloud if the bubbles will land in a new spot if they are blown from the top of the bale. Explore the texture of the straw with children, comparing it to other surfaces, such as grass, dirt, and gravel.

End

If outside time follows next in your daily routine, leave the bubble supplies and straw bales out for all children to enjoy. To help mark the end of small group-time, enlist the children's help in bringing out additional outdoor play equipment (bikes, balls, etc.).

Follow-up

1. Leave straw bales in the yard, encouraging children to move them around to different locations on the playground.

2. After a few days, use the straw from some of the bales in a scarecrow-stuffing small-group time. Arrange the scarecrows near the whole bales that are left outdoors, and put pumpkins of different sizes and shapes nearby.

Bales of straw scattered around the play yard provide a new jumping experience for children.

Soda-Bottle Bowling

Originating ideas

Several of the children in the classroom have parents who are members of a community bowling league and sometimes, the children go along. They occasionally talk about or re-enact these experiences with classmates and teachers.

Possible key experiences

Creative representation: *Pretending and role playing*

Language and literacy: *Describing objects, events, and relations*

Initiative and social relations: *Building relationships with children and adults*

Time: *Starting and stopping an action on signal*

Materials

✔ A big box of empty **2-liter plastic soda bottles** (approximately 30–35), partly filled with water or sand

✔ **Large rubber playground balls** and **tennis balls**

✔ For back-up: **Watering cans with long, thin spouts** for adding water to the soda bottles; **paper** and **small pencils** like those used to keep score at the bowling alley

Beginning

Meet with children in a large open space. Set up three or four soda bottles. Ask one of the children who has gone bowling to suggest where he or she might stand to roll a ball at the bowling pins. Once the child has taken a position, give him or her a ball to roll in the direction of the bottles. Then ask another child to help you set up more pins, and give that child a chance to "bowl."

Middle

Bring the rest of the bottles and balls out, and let children begin playing. Expect that some children will be interested in exploring the materials, simply rolling the bottles across the ground to see the water or sand move. Others will be interested in representing their experiences with bowling. Watch these children to see if they play alone, setting up their own bottles and knocking them down, or if they play cooperatively, perhaps taking turns at setting up "pins" and bowling. Some children may even count the number of bottles knocked over and request paper and pencils to write the scores down. Listen for any comments describing this experience or comparing it with real bowling to gain a better understanding of the amount of language children use to communicate with others.

End

When small-group time is almost over, tell children they have time for one more try at knocking over bottles. Put the bottles back in the large box, and enlist children's ideas about where they should be stored in the classroom. Before moving to the next part of your routine, ask children what their families do when their bowling games are over. Try to incorporate their ideas into your transition (for example, if they say they get in the car and drive home, pretend with children that you are driving as you move to the area of the classroom where the next activity will be held).

Follow-up

1. Ask parents if they can bring in some props from the bowling alley, such as score sheets and the special pencils used for keeping score. Some families may donate old bowling shoes or tournament shirts they no longer use. Add these materials to the classroom, stored near or with the soda bottle box.

2. Save a few of the partially filled soda bottles. At the next day's planning time, have children roll them into the area of their work time choice. Use a similar strategy at recall time.

Donald used cardboard tubes and a ball to act out what he saw at his parents' bowling league. This inspired teachers to plan the soda-bottle bowling activity.

Storytelling With Props

The preschool recently closed for two days during the Thanksgiving holiday. During the holiday break, many of the children traveled to see relatives or had relatives stay at their homes.

Possible key experiences

Creative representation: *Relating models, pictures, and photographs to real places and things*

Language and literacy: *Having fun with language—listening to stories and poems, making up stories and rhymes*

Classification: *Exploring and describing similarities, differences, and the attributes of things*

Seriation: *Comparing attributes (longer/shorter, bigger/smaller)*

Materials

✔ A collection of different-sized **pine cones, sea shells, rocks,** or **sticks** for each child

✔ **Inch-cube blocks** or **Cuisenaire rods**

✔ **Small toy vehicles**

✔ For back-up: **small toy people** or **teddy bear counters**

Beginning

Using your own set of materials, tell a short story about a school closed for two days. Use the different-sized pine cones, sea shells, rocks, or sticks to represent the family members of the children in the school, using the props to enact the story as you tell it. The story should describe how, during those two days, some of the children went on long car rides to visit people, others stayed at home with babysitters, and still others had guests come to their houses.

Middle

Watch the ways children manipulate materials. Expect that children will do various things with the pine cones, seashells, rocks, or sticks. For example, some children may arrange them according to size, others may scratch them with their fingernails to listen to the sounds created. Some children may combine the materials provided with other materials and use them to act out made-up stories about their own stay-at-home days. If they use the materials to enact stories, listen carefully and acknowledge feelings that arise ("So it was a little crowded in your bed and that was uncomfortable [or cozy]"). Ask questions when appropriate ("You ate a big meal? I wonder what you ate or liked best?").

End

Let children know when there are just a few minutes left in small-group time, asking them if there are things about their work that they would like to share with the others. Work with children to collect the materials and sort them into their containers.

Follow-up

1. Add books about family gatherings to the book area.
2. Use pine cones, seashells, rocks, or sticks as props the next day at planning or recall time. Encourage children to pretend their props are people who are talking about or enacting the child's plans or experiences.

Holiday Greetings

Originating ideas

To celebrate the December holidays, each year the preschool class takes a field trip to a local nursing home to bring holiday fruit baskets and poinsettia plants to the residents. During the visit, children and residents share a snack and, seated at long tables, work together on a large mural. Following this, they sing holiday songs together. This year, as the class discussed the upcoming field trip, three of the children in the group recalled last year's visit to the nursing home. One of these children asked if they could bring a special greeting card to give to the residents, "right before the singing starts."

Possible key experiences

Language and literacy: *Writing in various ways—drawing, scribbling, letterlike forms, invented spelling, conventional forms*

Initiative and social relations: *Being sensitive to the feelings, needs, and interests of others*

Materials

✔ The front panels from used **special occasion greeting cards** (families may be asked to donate these.)

✔ An assortment of different-colored **oak tag pieces** or other **heavy paper** to serve as backings for the card fronts

✔ **Glue**

✔ **Magic markers**

✔ For back-up: **clear tape**

Beginning

Ask the children who have been to the nursing home in the past to recall what happens there. Acknowledge other children's comments as they make them ("So you have a Grandma who lives in a hospital home"). Tell them you have brought materials for making cards to take to the nursing home. Explain that the group will give the cards, along with fruit baskets and plants, to the people who live at the home.

Middle

Watch as children create their own designs. Expect that some will spend time simply looking at the card faces, describing what they see and what they are reminded of. Others will attach the cards to the brightly colored oak tag and dictate messages for you to write on their cards. Other children will request markers to "write" their own messages using scribbles and letters.

End

As small-group time is about to end, put a big box in the middle of the table to hold the cards children will be taking to the nursing home. Expect that some children may want to take their cards home instead of giving them away. Permit this.

Follow-up

1. Put the materials in the art area so children can continue to create special occasion cards.

2. If there are extra child-made greeting cards that are not going to the nursing home, store them in a box near the greeting card materials as a reminder to children that they may choose to make more cards to add to those already inside. If occasions for card-sending arise, such as a child's illness or a person's birthday, point out the box and the materials to children.

Tissue Paper/Colored Cellophane Window Creations

Originating ideas

Recently, the preschoolers were invited to watch the choir in the neighborhood church rehearse for their annual Christmas program. While there, several children were especially interested in the stained glass windows, noting the colors of the glass and the designs and pictures created in each window by the arrangements of the small colored panes.

Possible key experiences

Language and literacy: *Describing objects, events, and relations*

Initiative and social relations: *Solving problems encountered in play*

Seriation: *Comparing attributes (longer/shorter, bigger/smaller)*

Materials

✔ Small, brightly colored pieces of **colored tissue paper** and **colored cellophane**

✔ Clear **tape**

✔ A large **window** or several **small windows** to attach tissue paper to

✔ For back-up: **white paper** (Some children may prefer to create designs on a horizontal surface or may find it difficult to work at a window that is crowded with other children.)

Beginning

Encourage children to recall some of the events of the previous day's trip by saying something like this: "Let's make up a story together about our visit to the church yesterday. I'll give the beginning, and everyone can take turns adding to the story. I'll start with this: 'Yesterday we went to hear the choir sing at the church, and while we were there we saw. . .'" When someone recalls the stained glass artwork, tell them you've brought materials they can use to make their own window designs. (If no one brings up the stained glass, prompt them by saying "Then we looked up at the windows and saw. . .")

Middle

Set containers of tissue paper and tape next to several of the windows in the room. As children get started, you may notice that some are interested in holding the different colored samples up to the windows to compare the colors and the changes that result when light shines through the paper. Other children may immediately start taping paper pieces to the windows, with some making random arrangements of paper pieces and others making patterns (for example, a row of red pieces followed by a row of yellow ones). As children work, acknowledge and describe their individual efforts.

End

When the group is nearly over, tell children that they have a few more minutes left to work. Wonder aloud what their designs will look like from the other side of the window. Listen carefully to their predictions and descriptions.

Follow-up

1. At the first opportunity to be outside after the group ends, be sure to look with children at the designs.
2. Add multicolored, translucent, plastic materials to your toy area so children can hold other materials up to the window to see the changes created by light.
3. As a planning/recall strategy, tape some colored tissue paper to the frames of a pair of eyeglasses, and have children look through the colored glasses into the area of their work time plans or experiences.

Ice Sculptures

Originating ideas

An ice sculpture show is held every year near the school neighborhood. During the show, large blocks of ice line the streets and artists use hammers and chisels to shape them into various objects. Several of the children in the class attended the event early on and watched the artists at work creating their sculptures. This small-group activity follows up on a walking field trip to the ice sculpture show, in which all the children in the classroom had a chance to view the finished ice sculptures. Teachers planned this activity to allow children to create their own ice sculptures. While many preschoolers may not be ready to use chisels and hammers, this center already had a large construction area, and the children were experienced with the tools used in the activity. In addition, most of the children were 4-year-olds who were mature enough to be successful with this activity.

Possible key experiences

Initiative and social relations: *Creating and experiencing collaborative play*

Space: *Changing the shape and arrangement of objects (wrapping, twisting, stretching, stacking, enclosing)*

Materials

✔ Several large **blocks of ice** (provide one large piece for each pair of children), made by filling plastic laundry tubs with water and freezing it overnight*

✔ The **bottoms from cardboard boxes** (to set the ice blocks inside to keep them from slipping as children work)

✔ **Mittens** for each child

✔ **Small chisels,** one per child, and **pounding tools** (one per child—these could be small hammers, mallets, or blocks)

✔ **Safety goggles** for each child

✔ For back-up: large **Styrofoam pieces** and **golf tees**

---

* For a younger, less experienced group of children, who may need more adult help to use construction tools safely and successfully, this activity could be modified by providing only one block of ice, to be used by two or three children, closely supervised by an adult. Meanwhile, the rest of the children could work with pounding tools, Styrofoam pieces, and golf tees, while awaiting their turn to work with ice.

Beginning

Meet around the large pieces of ice and ask the children who watched the ice sculptors at work to describe what the artists did to change the shape of the ice. Talk briefly with children about safety rules (wearing goggles, keeping hammers low, being aware of where others are standing) and include children in the discussion, encouraging them to ask questions and offer solutions.

Middle

When children all have goggles on, bring out the tools and let them begin chipping away at the ice. Stay close to children as they work and listen for their comments and observations. Some will notice how the ice crumbles as they chip at it, others will notice how the texture of the block changes from smooth to rough. Still others will notice how the box darkens as the ice chips melt inside it and it gets wet. Watch how children work with one another. For example, do they negotiate the sharing of materials and discuss where to chip next? For those children who use both the hammer and chisel at the same time, watch how they solve the problem of hammering while holding another tool. Be available to help if children experience social conflicts when working together, or if they experience difficulty chipping the ice (if this happens you may want to offer the Styrofoam pieces and golf tees, since these materials may be easier for some children to work with than the pieces of ice).

End

Collect any ice shavings or chips. Tell children you will store them in the freezer overnight to see what happens to them. Leave the ice blocks outside, and store tools and goggles in a box to bring back out the next day.

Follow-up

1. Bring ice scraps from the freezer to greeting time so children can comment on any changes they notice in the ice.
2. The next day at outside time, set out goggles and tools next to the sculptures in case children want to continue to reshape them.
3. Next to the ice-sculpting materials, add a box of accessories (hats, scarves, flowing fabric pieces) for children who wish to decorate their sculptures.

Making Hoagies (Submarine Sandwiches)

Originating ideas

Several of the children in the group have older siblings who participated in a hoagie-making project that was a fundraiser for a school trip. They took orders from the children's parents; over the weekend the actual hoagie-making and delivery took place.

Possible key experiences

Creative representation: *Recognizing objects by sight, sound, touch, taste, and smell*
Initiative and social relations: *Making and expressing choices, plans, and decisions*
Classification: *Exploring and describing similarities, differences, and the attributes of things*

Materials

✔ **Hoagie rolls,** one for each child (use the 6- to 8-in. size)
✔ An assortment of **sandwich fillings:** lunch **meats** (salami, ham, turkey), **cheese slices, sliced tomatoes, green peppers, shredded lettuce**
✔ Plastic **squeeze bottles** filled with **mustard, mayonnaise,** or **Italian dressing**
✔ **Paper plates, napkins**
✔ For back-up: **Paper and pencils** for order taking

Beginning

Ask those children who were involved to describe some of the events of the past weekend's hoagie-making project. When they are finished, tell the group that you have brought materials they can use to make their own hoagie sandwiches to eat at small-group time. Have children wash and dry their hands before using the materials.

Middle

Give each child a hoagie roll. Show children the other foods that are available for sandwich-making and watch as children decide what to put inside their rolls. Some children may decide to put mustard only in their sandwiches, while others will put everything on the table into theirs. Help children notice the differences in the choices they are making and the different textures, colors, and shapes of the ingredients. Listen for words like *squishy, juicy, crunchy, salty,* and *soggy.*

End

Bring the group to an end by having everyone eat their creations. Use the meal as another opportunity to talk about the choices children made and how their hoagies compare with the ones they ordered or helped make over the weekend. Bring storage containers for leftover food and, working with the children, clean the tables with warm, soapy water.

Follow-up

At large-group time, make up a song about the process of making a hoagie sandwich, using the children's ideas for the verses. Make up a variation on the song "Down Around the Corner at the Bakery Shop," as follows: "Down around the corner at the hoagie shop, there were hoagies with [child fills in ingredient] on top. Along came [insert child's name] and gobbled them up. Now there are no more hoagies at the hoagie shop."

Car Wash Day

Originating ideas

Several of the children, whose siblings are students at the local high school, recently participated in a car wash conducted to raise money for the high school band's annual trip. In the event, family members worked alongside the students as they washed cars.

Possible key experiences

Initiative and social relations: *Creating and experiencing collaborative play*

Movement: *Moving in nonlocomotor ways (anchored movement—bending, twisting, rocking, swinging one's arms)*

Materials

✔ Several **hoses** with nozzles attached

✔ Pails of **soapy water**

✔ Large, thick **sponges**

✔ **Dry towels**

✔ **Smocks** or bathing suits

✔ A dirty **car**

✔ For back-up: **other items to wash,** for example, **plastic wheeled vehicles, tables,** and **chairs**

Beginning

Have all the materials for car-washing available near or on top of the car. Meet with children around the car you will wash. Run your finger through a dirt spot and say, "Look what happens to my finger when I touch this dirty car." Encourage children to try the same thing. Ask the children who attended the car-washing fundraiser to describe what happened. When they are finished talking, pass out the sponges, place the soapy water pails within children's easy reach, and make hoses available.

Middle

Assist children as needed, helping them solve problems as they arise (for example, who will spray the water or how to spray it so it doesn't soak people who don't want to get wet).

Imitate and participate in children's ways of using materials, for instance, squeezing the sponges in the soapy water, rubbing the sponges on the car. Talk to children about whatever interests them or about the observations they are making (for example, about the bubbles in the pail and on their fingers, the puddles on the ground, the places where the car is covered in soap, and the different parts of the car—license plate, headlights, doors, and bumpers).

End

As small-group time comes to an end, tell children that they have time to wash one more spot before the final rinse. After the final rinse, pass out dry towels for each child and have them pick a wet spot on the car to dry. Help them notice the changes in the cloth and on the car as the cloth becomes damp. Finally, when they are done wiping their spot, ask them to rub their fingers on the car, as they did in the beginning, to see if their fingers come back dirty. Collect all pails, sponges, and other materials to take back inside.

Follow-up

1. The day after the small-group time, bring hoses, sponges, pails, and soap outside for additional water play.
2. Add soapy water, small sponges, and small plastic cars to the indoor water table.

Bike Riding

Originating ideas

The community recently sponsored a bicycle race through the downtown streets. Two children in the class entered in the age-5-and-under category and came to school wearing their prize ribbons. Their classmates listened with interest as the children described the race and the experience of "winning the gold medal."

Possible key experiences

Language and literacy: *Describing objects, events, and relations*

Movement: *Moving in locomotor ways (nonanchored movement—running, jumping, hopping, skipping, marching, climbing)*

Time: *Starting and stopping an action on signal*

Materials

✔ An assortment of **riding toys,** including **tricycles, big wheels,** and **wagons**

✔ **Signs** to mark the beginning and end of the bicycle course

✔ A **whistle**

✔ For back-up: **Additional whistles** or **bells**

Beginning

Gather outdoors with children on a surface suitable for riding. Ask children who attended the race to describe what happened. Ask for children's ideas on where to locate the signs for marking the beginning and end of the course.

Middle

Watch as children ride around, looking at the ways they maneuver around curves and one another. Observe their large muscle movements, watching how they start (by using their legs to push off or by pedaling) and stop (by stopping pedaling, by dragging their feet, or by running into things). Notice which children ride without comment, which children tell you of certain destinations they have in mind, and which children engage in real or pretend races, using the signs you have set out to mark the beginning and end of the course.

End

Warn children that the group is about to end, telling them you will blow a whistle in a few minutes when it is over. After you blow the whistle, have them park the riding toys for storage.

Follow-up

1. Bring a smaller version of the start sign, along with a planning map, to the planning table. Ask each child to place the start sign on the symbol for the interest area where their first work time plan will take place.

2. Bring the start and finish signs, the whistle, and the riding toys outdoors again to see if children use them in creating their own riding courses.

The opportunity to ride on a race course adds a new dimension to an always popular activity.

Garage Sale Play

Originating ideas

Garage sales are popular in the community surrounding the preschool, and many program children have had experiences with them. Several families of children in the program participate in a large neighborhood garage sale held every summer. For weeks before the event, families sort through clothing, household items, and garden tools, then spend two nights before the sale attaching price tags to each item. Materials are displayed on long tables in garages, and in baskets and on blankets in front of the houses.

Possible key experiences

Creative representation: *Imitating actions and sounds*

Language and literacy: *Writing in various ways—drawing, scribbling, letterlike forms, invented spelling, conventional forms*

Initiative and social relations: *Creating and experiencing collaborative play*

Classification: *Sorting and matching*

Materials

✔ Several **big boxes** (enough so that each group of three children has one big box) filled with a variety of materials donated by families (Include toys, clothing, books, and household items.)

✔ **Tables, baskets,** and/or **blankets** spread around the outside space

✔ For back-up: **Pocketbooks, paper bags, pretend money**

Beginning

Gather around the big boxes filled with donated materials and tell children that for small-group time today you will be having a school garage sale. Ask them to share their experiences with garage sales. If their families have participated in the big neighborhood sale or other garage sales, encourage children to recall some of the jobs their family members did to prepare for selling. Acknowledge what they say by repeating their words. ("You put stuff on tables and blankets, and people came to look at it"). Tell children that you will take the big boxes outside together so they can arrange their items for the sale.

Middle

Go outside and position yourself near the tables, baskets, and blankets. Observe the ways children sift through and sort materials. Some will put similar items together in a group, while others will group everything in one pile. Some children will start playing with the toys, while others will continue to pretend to be garage sale organizers. Imitate children's actions: if children are interested simply in playing with the materials, follow their lead; if their play involves acting out roles related to garage sales (such as customers or sellers), attempt to enter their play in a secondary role.

End

As the play ends, work with children to put everything back into the large boxes.

Follow-up

Bring a basket or blanket used in the small-group activity to the next planning time, and ask children to bring an item they will use in their work-time plans over to the basket or blanket. Then talk to the group about their plans for using that item.

Pizza Party Play

Originating ideas

The class has just taken a field trip to a local pizza establishment. While there, children got to handle the pizza dough, add toppings to the pizzas, and then eat the pizzas they had helped to make. The teachers decided to plan a small-group time that would give children an opportunity to recall and represent their experiences at the pizza parlor.

Possible key experiences

Creative representation: *Pretending and role playing*

Language and literacy: *Describing objects, events, and relations*

Initiative and social relations: *Creating and experiencing collaborative play*

Time: *Anticipating, remembering, and describing sequences of events*

Materials

✔ **Play-Doh**

✔ **Pizza pans**

✔ **Pizza cutters, rolling pins**

✔ **Pizza boxes** (small-, medium-, and large-sized)

✔ **Potholders, aprons**

✔ For back-up: **paper plates** and **paper napkins**

Beginning

Give each child a sizeable chunk of Play-Doh. Put additional materials on the table, telling children that the materials remind you of the trip you all took to the pizza shop. Tell children they can use the materials with their dough.

Middle

As you watch the ways the children approach the materials available to them, begin working your Play-Doh.

After the pizza-making small-group time, Erica (above) and Frances (right) expanded their pretend cooking with Play-Doh.

Listen for conversations children may have about their trip to the pizza shop. Involve yourself in their play by imitating their actions (for example, rolling, flattening, or squeezing your Play-Doh or pressing it into a pizza pan as you see them doing) and by making comments about their work ("I see you filled a big box with Play-Doh"). Occasionally add details to your work that they might not have considered on their own (for example, tearing off tiny balls, putting them on top of a flattened piece of dough and calling them "mushrooms"). Watch to see if children assign roles to each other ("Sit down—I'll bring the pizza to you, like at the restaurant") or play in other cooperative ways. Observe any cooperative play closely, noticing such things as how children respond when others initiate interactions with them and how long a group can sustain their cooperative play.

End

Warn children that small-group time is about to end, connecting what you say with what you see children doing (for example, "You have time to put two more pepperoni pieces on the pie" or "You can cut three more slices"). Return the Play-Doh to the sealed containers, then put the containers and the other pizza-making materials in a heavy cardboard box labeled with pictures taken during the field trip to the pizza shop.

Follow-up

1. Store the box of pizza props in the block or house area.
2. As a planning or recall strategy, draw a circle inside one of the cardboard pizza boxes, and draw an interest area symbol in each slice section. Give the children a pizza cutter to "slice" the area they plan to work in (or did work in) at work time.

Kindergarten Visit

Originating ideas

Each year, as the preschool year ends, the class takes a field trip to the local public school where most of the children will attend kindergarten. The field trip includes visits to a kindergarten classroom, the auditorium, and the playground, and a chance to sit inside the yellow school bus.

Possible key experiences

Creative representation: *Making models out of clay, blocks, and other materials*
Language and literacy: *Talking with others about personally meaningful experiences*
Time: *Anticipating, remembering, and describing sequences of events*

Materials

✔ A **basket** for each child filled with **small plastic people** and **inch-cube blocks**
✔ Several **toy yellow school buses**
✔ For back-up: **Cuisenaire rods**

Beginning

Start by "driving" a yellow school bus filled with plastic people around the table. Tell them you are the bus driver who is taking them for their first day of kindergarten, and wonder out loud what might happen there. Acknowledge children's responses to your question, then pass out a basket of materials to each child.

Middle

Watch as children work with the materials. Some children may use them to represent the playground, the classroom, or the auditorium they visited. Some of these children may use the plastic people as props as they talk about the things they remembered from the trip, so listen carefully and enter their conversations when invited. When appropriate, enact the roles of the people they encountered on the trip to the school (the bus driver, the classroom teacher, some of the students in the hallways), and when interacting with the children, use some of the words you heard those people use. Other children may use the blocks and people to stack and build in ways that are unconnected to the trip. Still others may only be interested in filling the school bus with people and driving it around the floor. Be sensitive to issues children raise about the field trip experience, acknowledging any feelings they may share through their role play, rather than downplaying their importance (for example, say "A new school is scary; it's hard to meet new people" instead of "Don't worry—it will be fun").

End

Tell children that it is almost time to get on the bus and drive back so that you can get to the preschool in time for large-group time (or whatever is the next event in your daily routine). See if they say goodbye to any of the characters from their role play or if they willingly begin putting away materials. Go around the table with the small school bus as you did in the beginning of the activity, pretending to pick up each child. As children take their turns "getting on the bus," have them refill their baskets with materials, and encourage them to talk about what they thought was their favorite thing about playing with the toys.

Follow-up

1. Use the bus the next day at planning time to follow the "bus driver" (the person planning) to the area he or she will work in.

2. At recall time, set up pairs of chairs side by side, and ask children to take seats with a partner and to pretend they are in a school bus. Then ask them to tell their partners what they did at school that day.

Firework Celebration Paintings

Children created a "fireworks explosion" by scattering toys at work time. To offer a constructive alternative to children, teachers later planned this small-group activity in which children can create another kind of fireworks.

Originating ideas

Fourth of July fireworks are an annual event in the community, and several children in the group recently attended this year's display. At snack time the teachers have overheard these children talking about going to the fireworks, and at work time they have observed the children throwing small toys from the toy area into the air and calling them "fireworks."

Possible key experiences

Creative representation: *Drawing and painting*

Language and literacy: *Talking with others about personally meaningful experiences*

Materials

✔ A mixture of two parts **paint** with one part **glue** and two drops of **water** in individual **cups** or **squeeze bottles**

✔ **Construction paper**

✔ **Glitter**

✔ **Straws**

✔ For back-up: **paintbrushes** or **cotton swabs**

Beginning

Ask the children at your table who attended the fireworks display to describe, in their own words, what they remember about what they saw. Invite other children to share any experiences they've had with fireworks. When the conversation winds down, tell children you've brought materials they can use to make their own fireworks pictures. Squeeze some of the paint-glue mixture in the center of a piece of paper; then hold up a straw and ask children to predict what might happen if you blew through the straw at the paint on the paper. Listen to their ideas about what they think will happen. Then tell children they can use the paint, paper, straws, and glitter to make their own colorful pictures.

Middle

As children begin working with materials, work alongside them making your own picture. Use your straw to blow paint across the paper, then imitate the ways you see other children using their straws. Some may use them for dipping and spreading the paint; others, for simple stirring of paint inside their individual cups. Sprinkle glitter on your paper, making comments that include some of the children's own words about their fireworks experiences ("Wow! When I do that it looks like the sparkles in the sky!"). Be available to listen to children's explanations of their work or their comments about their trip to see the fireworks. Note how accurate and detailed their descriptions are.

End

Ask children to put away their pictures in the space you use for drying children's artwork, to return unused materials to their storage areas, and to help clean the tabletops. Make a game of the transition to the next part of the daily routine by asking children to pretend they are "fireworks exploding in the sky" as they move to the space for the next activity.

Follow-up

1. Add containers of glitter and the paint-glue mixture to the art area.

2. Bring confetti to outside time, encouraging children to toss it in the air to make a different kind of "fireworks."

Outdoor Carnival

Originating ideas

Many of the children in the group attended a local carnival. During snack time for several days afterwards, the teachers observed children talking about the carnival rides they went on and the food they bought with their carnival tickets. The teachers have decided to combine snack time with small-group time for this outdoor carnival experience.

Possible key experiences

Initiative and social relations: *Making and expressing choices, plans, and decisions*
Number: *Counting objects*

Materials

✔ **Materials to set up a minimum of five activity stations** in an outdoor play space or a large indoor space (For example, you could have a **tricycle loop,** a **jumping spot,** a **bingo marker/ construction paper art activity,** a **beanbag toss,** and a **snack stand** serving something similar to the refreshments sold at the carnival—lemonade and popcorn.)

✔ **Cardboard boxes** to set at the entrance to each station (Cut a slot in the top of each one and label the box with a picture of the number of tickets needed to play the game—We recommend having some activities cost one ticket, others two, and others, three.)

✔ A **roll of tickets** for each child

✔ Back-up materials: **toy cash registers** and **play money**

Note: If possible, arrange to have parent volunteers help with this activity.

Beginning

Meet at the normal small-group spot and tell children that you will be having an outdoor carnival. Explain that during the carnival they will be able to ride tricycles, jump, make a picture, throw a beanbag, and/or have a snack. Set out the tickets, and show them a box with a ticket number on the outside. Explain that before they do their activity, they may pay for their choice with a ticket or tickets. Pass out their tickets, and let them put one in the ticket box before going outside.

Middle

Choose activities side by side with children, watching to see how children make choices. (For example, do they scan the whole environment first or do they rush right to a certain activity?) Notice which children put tickets in the slotted boxes and whether they put the suggested number of tickets in the box. (Some children may randomly put a pile of tickets in each slot without counting them, while others may ignore the tickets completely.) See what happens when or if children run out of tickets. Do they come back to an adult and ask for more, continue to play without tickets, or ask their friends if they will share?

End

Give a 5-minute warning near the end of small-group time, ringing a bell to alert children to your announcement. Explain at this time that after the carnival everyone will be meeting back inside to talk about their favorite part of the carnival. Then have everyone meet again at the small-group table, and encourage them to describe or draw a picture of their experiences.

Follow-up

1. Leave the boxes and materials outside for a few days so children can re-create their experiences over a period of time.

2. Use tickets (in the shape of the children's symbols) the next day at planning time as the children share their work time plans. Prepare shoe boxes labeled with the interest area symbols and cut slots in the lids. Ask children to put their ticket in the box representing the area they will work in.

Carnival Boat Play

Originating ideas

Children recently took a field trip to the annual community fair. During the field trip they watched carnival workers set up amusement park rides. After the field trip, the adult heard children talking about the "ride with the boats in the water," with one of the children repeating the language of the carnival workers: "Step right up and into the boat."

Possible key experiences

Creative representation: *Relating models, pictures, and photographs to real places and things*
Movement: *Describing movement*

Materials

✔ A **sand and water table** filled with **water** and at least **two other large plastic tubs** filled with water (Laundry-tub-sized containers with handles on each side work well.)

✔ A collection of **plastic boats,** enough so that each child has at least two

✔ An **assortment of objects** small enough to fit inside the boats, for example, **small toy people** and **teddy bear or dinosaur counters**

✔ For back-up: A variety of different-sized **plastic containers** suitable for filling and emptying

Beginning

Meet around the table and tell a short story, using the materials listed above as props. In the story, describe a carnival where children are excited to see boats floating in the water ready to take them on a ride, but are wondering how to get into the boats without getting wet. Then give each child a paper bag containing the small boats and toy figures. Have children gather around the sand and water table and the separate plastic tubs.

Middle

Watch to see if any of the children's play or conversation includes details about what the carnival worker who operated the boat ride said. Observe the ways children fit the toy figures inside their boats and the movements they use to propel the boats through the water. Listen for any language from children about how their boats are moving, what they remember from the field trip, or details from the short story you told at the opening of small group. Sail your own boat through the water and respond if children invite you to join in their actions.

End

Put on a carnival-style musical selection, such as "Entertainer" from High/Scope's *Rhythmically Moving 5* recording. Tell children that when the music stops, the carnival ride will be over and it will be time to put the boats and riders away. As the music ends, observe to see which children put away materials without adult or child prompting.

Follow-up

1. Bring one boat and one toy figure to the planning table the day after your small-group time. Draw the symbols of your classroom areas on a round blue sheet of paper, and ask the child who is sharing work time plans to float the boat to the area he or she will work in.

2. Add to the book area *Big Bird Joins the Carnival* by Cathi Rosenberg-Turow.

The day after the carnival boat activity, Christopher can't wait to fill up the water table.

Farm Animal Recall

This class trip to the 4-H Club farm show is the inspiration for a small-group time in which adults provide photos of the trip, toy farm animals, materials for building animal stalls, and tiny combs and brushes.

Originating ideas

This activity was planned to follow up on a trip to the community's 4-H club farm show, an event where children display animals they have raised: llamas, goats, sheep, pigs, rabbits, chickens, horses, cows, ducks, and geese. While at the show, teachers and children had the opportunity to observe some of the child owners grooming their animals and to see the prize ribbons, which were posted on the stalls and cages.

Possible key experiences

Creative representation: *Relating models, pictures, and photographs to real places and things*

Language and literacy: *Describing objects, events, and relations*

Time: *Anticipating, remembering, and describing sequences of events*

✔ **Photographs** taken while attending the show

✔ **Toy farm animals**

✔ A variety of **building materials** for creating stalls and cages (Popsicle sticks, Lincoln Logs, or Cuisenaire rods are possibilities.)

✔ **Tiny combs or brushes** for pretending to groom the animals

✔ For back-up: a selection of different-colored **ribbons** to represent the prizes

Beginning

Bring pictures from the field trip to the group. Look through the photos together, encouraging children to talk about what they remember from the field trip. After a few minutes have passed, bring the additional materials to the attention of the children, saying "I have some materials that are like the ones we saw on the field trip. See what you can do with them."

Middle

Listen as children recall and re-create the trip in their own ways. Some children may remain interested in the photographs, using words to describe their experiences. Others may choose to use the materials available to recall the trip through their actions. Note which children recall the animal grooming and caretaking they observed, and which focus on the showing of the animals or the awarding of the prize ribbons. Some children may be primarily interested in creating cages to house the animals. As you observe children re-enacting their experiences, notice how many details of the original farm show children include in their re-enactments. Expect that some children may simply play with the materials in ways that do not relate to the show.

End

Ask children to describe for the group the favorite parts of their work with the animals or of the field trip. Working with the children, sort the animals, building materials, and grooming supplies into separate containers.

Follow-up

1. This strategy uses an animal cube you have made by taping photos of six of the animals you saw at the show onto each surface of a plastic or wooden cube. At large-group time, ask children to take turns throwing the cube into the middle of the circle. Make up a group story about the animal whose picture is face-up when the cube stops.

2. Put the small-group materials in the toy area to offer new play choices for future work times.

3. At planning time the day after the small-group activity, construct a cage or stall on the planning table and set toy animals inside it. Ask children to take turns "letting an animal out of its cage" and carrying it to the area of their work time choice.

Materials Index

A

Abiyoyo, retold by
   Pete Seeger, 54
Acorns, 74
Aluminum foil, 58, 102
Angry Arthur, by
   Hiawyn Oram, 152
Animals, toy farm, 224
Apples, 76
Aprons, 214

B

Bagels, 48
Bags, paper, 144
Balloons, 64
Balls
   playground, 196
   soft foam, 80, 158, 160,
      180
Barrettes, 26
Bats, soft foam, 64, 160
Beans, 54, 128
Blankets, 138, 180, 212
Blocks
   Bristle, 72
   inch-cube, 80, 130, 198,
      216
Boats, plastic, 222
Books, 44, 54, 98, 106, 126,
   130, 152, 166, 190
Bowls, 80, 98
Boxes, 20, 88, 212, 220
Branches, 38
Bubble solution, 156, 194
Bubble wands, 156, 194
Buses, toy, 216

C

Camera, 176
Cardboard
   boxes, 20, 88, 128, 130,
      212, 220,
   pieces, 30, 38, 56, 58,
      68, 104, 116, 136, 170,
      200
   tubes, 54, 64, 168

Cards
   greeting, 200
   playing, 92, 128
Carpet squares, 166, 170
Car, 208
Cars, plastic, 148, 198
Cellophane, colored, 202
Cereal, instant hot, 98
Chalk, 118
Cheese slices, 206
Chisels, 204
Chopsticks, 80, 154
Clothes, 50, 190
Clothesline, 50, 158
Clothespins, 50, 80
Coffee cans, 22
Coffee filter paper, 58
Color Zoo, by
   Lois Ehlert, 126
Condiments, sandwich, 206
Cookie-cutters, 150
Corks, 72
Cotton balls, 32, 94, 170
Cotton swabs, 32, 94, 170
Crayons, 24, 84, 102, 104,
   126, 136
Cream cheese, 48
Crystal Climbers, 92
Cuisenaire rods, 130, 198,
   216, 224

D

Dental floss, 132
Dogs, 36
Dolls, 138
Dowels, 138

E

Egg cartons, 128
Eggshells, 60
Egg timer, 98
Elastic, 140
Eyedroppers, 58, 90
Eyeglass frames, 26

F

Fabric
   cotton strips, 138
   flannel, 170, 182
   muslin, 24
   thick cotton, 62
   white, 100
Feathers, 112
Flannel boards, 182
Flashlight, 140
Flowers, 24
Fly swatter, 156, 194
Food coloring, 28, 58, 90,
   156

G

Glitter, 26, 54, 150, 170,
   178, 186, 218
Glitter glue, 104
Glue, 26, 28, 30, 38, 52, 54,
   56, 58, 60, 68, 78, 84, 106,
   142, 144, 150, 152, 164,
   168, 170, 178, 186, 188,
   200
Glue brushes, 56, 58, 170,
   186
Glue sticks, 60
*Goldilocks and the
   Three Bears,* 98
Golf tees, 120, 204

H

Hair curlers, 72
Hammers, 24, 120, 204
Hand mirrors, 26, 34, 132
Horse chestnuts, 74

I

Ice, 42, 78, 98, 204
Ink pads, 44, 100
Ink stamps, 44, 100

L

Ladles, 94, 98, 146
Laundry tubs, 50
Leaves, 184, 188
Lizards, plastic, 66

M

Magazines, 152
Magnets, 110
Magnifying glasses, 74, 178,
   192
Marbles, 88
Markers, 24, 30, 44, 54, 56,
   60, 62, 68, 80, 96, 100,
   102, 104, 114, 118, 126,
   136, 152, 182, 186, 200
Meats, lunch, 206
Medical supplies, 138
Microwave oven, 98
Mirrors, 26, 34, 132
Mittens, 204
Muffin tins, 32
Music, recordings 96, 154,
   162, 166, 172

N

Napkins, 48
Noodles, 30
Nuts and bolts, 116

P

Packing peanuts, 78
Pails, 26, 108, 146, 148
Paint, tempera, 32, 46, 62,
   68, 72, 86, 88, 114, 144,
   168, 178, 188
Paintbrushes
   easel, 22, 46, 62, 68, 88,
      114, 142, 144, 148,
      168, 178, 218
   housepaint, 22, 46
   watercolor, 22, 46
Paint rollers, 22
Paint trays, 22
Paper
   construction, 46, 52,
      58, 60, 72, 78, 86, 88,
      100, 104, 118, 126,
      148, 152, 178, 182,
      218
   scraps, 114, 134, 144,
      170
   strips, 44, 84, 112
   waxed, 58

About the Author

Michelle Graves, a writer and educational
consultant at High/Scope Educational
Research Foundation, has designed and conducted training workshops and long-term
training projects for teachers, teacher-trainers, and educational administrators in a wide
range of educational settings, including Head Start and other preschool, child care, and
special education programs. She is the author of *Daily Planning Around the Key Experiences*
and *Planning Around Children's Interests,* two previous books in *The Teacher's Idea Book*
series. Graves is also the script author/coproducer of the videotape *Supporting Children's
Active Learning: Teaching Strategies for Diverse Settings* and the *Small-Group Times*
videotape series, both from High/Scope Press. Formerly, Graves directed a child care
center for employees of the Veterans Administration Medical Center, Ann Arbor, Michigan,
a program serving families with children aged six weeks to five years. She has extensive early
childhood teaching experience. In addition to the High/Scope Demonstration Preschool,
Graves has taught in public and private child care and special education programs.

Related High/Scope® Resources

The Teacher's Idea Books

Making the Most of Plan-Do-Review: The Teacher's Idea Book 5

Children in High/Scope settings learn how to plan their activities, accomplish their goals, solve unexpected problems, make necessary changes to their original plans, and reflect on the outcomes of their actions. These are valuable skills they will use throughout life. High/Scope's daily plan-do-review process makes it all happen. This book provides a blueprint for successful implementation. Includes support strategies, practical tips and suggestions, tried-and-true games and experiences for children, answers to frequently asked questions, real-life examples, sample notes to parents, a parent workshop plan, and planning and recall sheets to use with children.

BK-P1152 $25.95

N. Vogel. Soft cover, photos, 270 pages. February 2001. 1-57379-086-9

Planning Around Children's Interests: The Teacher's Idea Book 2

Like the others, the second book in this popular High/Scope® series is filled with practical teaching strategies and actual classroom examples of teacher-child interactions. All new, up-to-date, and fun for all, the ideas draw on children's interests as a rich resource for curriculum planning. An essential handbook for dedicated professionals.

BK-P1106 $25.95

M. Graves. Soft cover, photos, 171 pages, 1996. 1-57379-019-2.

Buy all 5 & SAVE!
Special package price
BK-P1161SET
$110.00

The Essential Parent Workshop Resource: The Teacher's Idea Book 4

If you are interested in presenting workshops for parents of preschoolers, you will be delighted with this collection of 30 original workshops. Presenters will find it easy to follow the workshop format, which includes intended goals, a list of necessary materials, an introduction and interactive opening activity, central ideas for discussion, scenarios for reflection and application of ideas, and follow-up plans that encourage parents to apply the information at home. Packed with handouts and charts, it's all you need for practical, dynamic parent workshops.

BK-P1137 $25.95

M. Graves. Soft cover, photos, 180 pages, 2000. 1-57379-018-4.

Daily Planning Around the Key Experiences: The Teacher's Idea Book 1

Make each part of the daily routine a useful and focused learning experience for preschoolers and kindergartners with the practical, creative suggestions in this handbook. Provides specific ideas for each part of the daily routine, including suggested materials, questioning techniques, and ideas for small- and large-group activities.

BK-P1076 $19.95

M. Graves. Soft cover, 87 pages, 1989. 0-931114-80-2.

High/Scope's Preschool Manual and Study Guide

Educating Young Children: Active Learning Practices for Preschool and Child Care Programs

Written for early childhood practitioners and students, this manual presents essential strategies adults can use to make active learning a reality in their programs. Describes key components of the adult's role: planning the physical setting and establishing a consistent daily routine; creating a positive social climate; and using High/Scope's 58 key experiences to understand and support young children. Other topics include family involvement, daily team planning, creating interest areas, choosing appropriate materials, the plan-do-review process, small- and large-group times. Offers numerous anecdotes, photographs, illustrations, real-life scenarios, and practical suggestions for adults. Reflects High/Scope's current research findings and over 30 years of experience.

BK-P1111 $39.95

M. Hohmann & D. P. Weikart. Soft cover, lavishly illustrated, 560 pages, 1995. 0-929816-91-9.

Now available in Spanish. BK-L1016 $39.95

A Study Guide to Educating Young Children: Exercises for Adult Learners

The study guide you've been waiting for—a must-have workbook for High/Scope's latest preschool manual! Designed for early childhood college courses, inservice training, and independent study. Will increase your confidence and competence in using the High/Scope® Preschool Curriculum. Contains active learning exercises exploring the content of the manual in depth. Chapter topics parallel *EYC's*. Abundant, interactive exercises include hands-on exploration of materials, child studies, analysis of photos and scenarios in *EYC,* recollection and reflection about curriculum topics, trying out support strategies, and making implementation plans.

BK-P1117 $15.95

M. Hohmann. Soft cover, 275 pages, 1997. 1-57379-065-6.

To order these or any other High/Scope® products, contact High/Scope® Press: phone (800)40-PRESS fax (800)442-4FAX
To see a full listing of High/Scope® preschool products, visit our Web site: www.highscope.org